BLUEPRINTS FOR A SPARKLING TOMORROW

BLUEPRINTS FOR A SPARKLING TOMORROW

Thoughts on Reclaiming the American Dream

(revised and expanded edition)

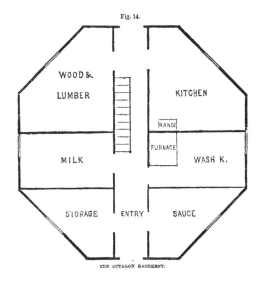

Fig. 14.

THE OCTAGON BASEMENT.

By NATHAN J. ROBINSON
and OREN NIMNI

DEMILUNE PRESS
AUSTIN | FT. WORTH | NEW HAVEN

*to the veldt and all those who have expended
their energies in attempting to save her*

Sycophantic Palms Press
Boston – Seattle – St. Louis
Copyright © 1965, 2010, 2015 by
Oren Nimni and Nathan J. Robinson
All rights reserved
Distributed in Canada by Nimni & Associates
Printed in the United States of America
First edition published in 2010 by Borometric Language Press

Library of Congress Catalog-in-Publication Data

Nimni, Oren and Nathan J. Robinson
    Blueprints for a sparkling tomorrow: thoughts on reclaiming
the Armenian dream / Oren Nimni and Nathan J. Robinson
  2 rev. and expanded ed.
      p. cm.
  Includes index.
  ISBN-13: 9780692479810
  ISBN-10: 0692479813
 1. Diffusion of innovations. 2. Information society. 3. Globalization
–Economic aspects. 4. Globalization – Social aspects. 1. Title.
HM846.F74 2010
303.48'33-dc22

TABLE OF CONTENTS
(sparse)

TABLE OF CONTENTS
(well-fleshed)

IV. GROWTHS...121

(that which does not yet exist, but must and will)

This Table of Contents is purely aspirational.

BLUEPRINTS FOR A SPARKLING TOMORROW

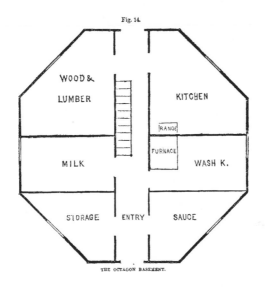

Fig. 14.

THE OCTAGON BASEMENT.

By NATHAN J. ROBINSON
and OREN NIMNI

Thus then intelligence, or in other words plain unsophisticated reason, will consider the various sentiments and actions which now create misery in society, will patiently trace the cause whence those sentiments and actions proceed, and immediately apply the proper remedies to remove them.
- Robert Owen, <u>A New View of Society</u>, 1816.

Preface
THE VINDICATION OF VARIOUS HORSE-BASED THEORIES & A NOTE ON THE REVISED TEXT

Several years ago we condemned the horse. In fact, our work up until this point has been littered with references to horses and horsing.[1] By contrast with certain discreditable academics from certain former Soviet satellite states, we have never been fooled by the horse's wistful eyes and sympathetic grin. We have built forests-worth of CV-pages on horse-based articles and public presentations, and denounced horses from every podium to which we have ever been invited or from which we have been removed.

Yet still, the horse persists. People ride horses, people videotape themselves being coddled by horses, people *intentionally purchase sugarcubes from supermarkets and go ahead and hand them to horses.* None of this would have happened had our scholarly work been given its due, and we must reluctantly concede that the continuance of this behavior suggests our journal articles are not being read. Horses, it seems, are here to stay.

But though we have come to accept the inevitable eternity of this niggling inconvenience, we feel obliged to point out that since the publication of our last major treatise, our suppositions have been largely vindicated, and the miseries of a horse-based life have been exposed by consequence. Consider the following:

> *Bellevue, Iowa — Two runaway horses trampled onlookers, including children collecting candy, at a Fourth of July parade in this Mississippi River town Sunday, killing one person and injuring 23 others.*
>
> "Runaway horses kill one, injure 23 at Iowa parade,"
> The Los Angeles Times, July 5th, 2010.

By now the dangers should be only too clear! These Iowans now know what we have for aeons, namely that horses and reason can not

---

1   See: Nimni, Oren and Nathan J. Robinson "Horses & the Subjugation of Vice," *Fordham Law Review,* May 1997.

coexist without ruckus (to speak logically: horses ∧ humans ⇒ruckus). The Iowa horse-murder is the capstone to our theorizing, and we cannot imagine a more perfect instance of having told them so.

Our pet theory does, of course, run directly contrary to the argument spewed by Professor Alan Dershowitz in his third book of 2003, *The Case Against Horses' Enemies*. However, we see no need to give this criticism serious attention, as Dershowitz's shabby grad-student-penned demi-pamphlet has already been given both wide and serious mockery in Equine Studies journals across the country. Furthermore, Dershowitz is said to have received his current Harvard sinecure largely thanks to his mother's vigorous lobbying of the tenure committee.

Hence we have no time for Dershowitzian horse-folly. What would Mr. ("Prof.") Dershowitz have to say to 24 injured and/or dead Bellevuians, if he were asked to repeat his calumnies in earshot of his victims? We do not know, but we can only suspect that when confronted with the human face of his crimes, he would cower and/or flee.

Practically speaking, can the horse be exorcised? We do not know, nor do we intend to find out. It is a brief life, even for the longest among us, and to spend it dithering with horses is to spend it idly. The design and implementation of workable horse-reduction policies is beyond the scope of our ignorance. With this, then, we dispose of the subject, never to return to it except fleetingly throughout the remainder of the text.

A NOTE ON THE REVISED TEXT

We first spoke these *Blueprints* in the autumn of 2010, a somewhat poisonous time in American life during which reason and good sense had been temporarily suspended for reasons of national security. As a result, the published text seemed uncobbled and disjoined to the unfamiliar reader. The work did receive strong early reviews in the pages of the *Sensible Utopian* and the *Yale Review of Sorry Excuses For...* (classified as "a sorry excuse for a badly-written leaflet," ergo a well-written leaflet?). But after a blistering disparagement by a certain quasi-notorious 7th Circuit jurist in the pages of the *New Republic* (part of man and magazine's three-part series "An Esteemed Judge Scurrilously Damages The Reputations of Five Leading Legal Scholars"), we had our forebrains hammered by the realization that textual finessings and caressings would be due.

Friends were skeptical. "How can you revisit what was never ineffable?" was the common cry, understandable after so much water had been passed. For some years we dandled and buoyed the text like a casket at a wedding, uncertain and free of oomph. We then mind-changed, and intended fully to split the pea into three volumes, Volume I being the original *Blueprints*, with Volumes II and III to be entitled *Dimension of Communitopia: An Exercise in Sane Living*, and *The Human Disease: Its Cause and Its Cure*. A tripartite trilogy of prophetic texts, followed perhaps by a sensibly-priced omnibus edition entitled *The Collected Nimni-Robinson Lectures on Social Tension & Decay*. This, too, however, made a small part of us continuously uneasy.

Alas, eventually we hit upon the ticket. New *Blueprints* would be written, with the original *Blueprints* republished as *The Human Disease*. Then when *Dimensions* were written later, they too could initially be published as *Blueprints*, with the second volume (this one) retitled *Dimensions* until the republishing of the third. The plan made so much sense that our publisher wept.

Yet the scheme has undergone one further mutation since this final formula, one critical to understanding what it is that you now hold. The third volume was canceled and folded into the second, with all other components remaining the same, except that instead of being a new second volume, *Blueprints* was to be a heavily-revised edition of its original namesake. Thus, the present book combines highlights of the early *Blueprints* (with various ideological corrections) with the material that was ultimately to comprise volumes II and III as well as the sensibly-priced paperback omnibus. In this manner is wisdom transformed into ink and pages.

INTENTIONS

On the rear jacket of the first work we laid out our intention for the series as follows: to diagnose the human disease. Upon re-reading the body of our former *corps*, however, we realized that we largely lost sight of this initial mission over the course of our blusterings.

We also subscribed to a flawed principle, which requires acknowledgment hither. We believed that one of the primary causes of suffering is the recognition of suffering itself. By choosing to identify a problem, we supposed, one heightens the experience of it, thereby compounding its problemitude. A self-fulfilling prophesy.[2] Laboring under this fatally skewed reasoning, we left answers to the question "What plagues mankind?" intentionally vague, supposing such vaguery to be the answer best suited to alleviating such plagues. It would get mankind thinking!

But it didn't, and doesn't. The human disease still gnaws at the organs of even the saintliest.

We wish, then, to clamber once again upon our now-somewhat mealworm-laden soapbox, and finally to fulfill our divine charge. We will, over the course of the ensuing pages and paginettes, offer both Diagnosis and Cure for what ails us (and by extension, you).

---

2   To give a parallel example of a self-fulfilling prophesy: one of us recently announced to a group of assembled colleagues that he had lost his glasses. When those colleagues shouted "No, you haven't, they're on your head!" in unison, the esteemed professor was forced into such a defensive position that he *could not acknowledge* the veracity of the statements, in spite of their obvious truth. His denials grew in assertiveness in equal proportion to the vigor of their accusations, so that by *stating the glasses were lost* he led himself to *not finding the glasses*. This and a thousand similar incidents have prompted us to discard the notion of the self-fulfilling prophesy entirely in our more recent works. Ego stands in the way of the recognition of truth.

Introductory Segment

"Convinced that all previous philosophers had done nothing to solve the problem of human happiness, Fourier ignored them and their 400,000 nonsensical volumes. Relying on his own imaginative powers and a smattering of scientific knowledge, he spent his remaining years elaborating his theory...and setting down on paper a minutely detailed account of the fulfilling life that man was destined to lead in the ideal realm which he called Harmony."
- <u>The Utopian Vision of Charles Fourier</u>, p. 2

"Comforting as the hug may be, principals across the country have clamped down. "Touching and physical contact is very dangerous territory," said Noreen Hajinlian, the principal of George G. White School, a junior high school in Hillsdale, N.J., who banned hugging two years ago. "It was needless hugging — they are in the hallways before they go to class. It wasn't a greeting. It was happening all day." – <u>The New York Times</u>, May 28, 2009 "For

Teenagers, Hello Means 'How About a Hug?'"

Contention: A world in which the preceding string of words can be truthful is a world which requires drastic and immediate adjustment.

We propose to initiate this adjustment through the ensuing text, and to liberate human beings from the circumstances that enslave them. We will lay out the problems in the order of their appearance. Exploitation! Isolation! Alienation! Desperation! The Criminalization of the Hug! All of these are universally acknowledged, yet the terrible blight they inflict upon us remains unaddressed.

No longer! The slim volume resting in your palms represents the culmination of twenty-five years of research and reflection, distilled and transmogrified for consumption by a general audience. Utilizing the full power of the Science of Present and Future Beings, we will train the willing and unwilling alike to resist the representatives of external coercion.

Five centuries of humbug may finally come to an end, as we present you with Blueprints for a Sparkling Tomorrow.

THE DYNAMIC NATURE OF THE BLUEPRINTS

In spite of appearances to the contrary, the Blueprints are not sacred scrolls to be ceremonialized. They are a dynamic creation, shape-shifting and mind-morphing as they saunter through Time. We do not presume to know what application of the blueprints is "going to look like," and pay no attention to the dogmatism of conventional architectural teachings. If we advocate triangulated tensegrity in one paragraph, we may invert ourselves and argue for the superiority of bilinear intensitegrity in the succeeding paragraph. The blueprints go where they are needed and are used according to the user's particular daily dosage requirements. If you wish to cross out or replace any portion of the blueprints, you are welcome to do so.

This work is intended not as a Universal Declaration or Cyborg Manifesto, but rather as a kind of *putty for the eyes* that works on any level upon which one cares to place it. This book is not intended to be a manual of archaeology, still less of the history of science. It is meant to be readable to those who are not concerned with the detailed problems about which specialists argue heatedly. It is merely a short and lucid study of the origin and progress of man from earliest recorded history to modern times, and a set of nebulous prescriptions for the improvement of history's monorail or commuter-bus-system.

Books accidentally dropped in storm drains will not be replaced.

WHAT YOU WILL BE ABLE TO DO AFTER READING THE BLUEPRINTS TWICE DAILY FOR A FORTNIGHT

We are neither promissors nor prelates, and so do not guarantee specific levels of Results or Consequences from the perusal of our tome. It will not Open Your Love Windows or Wax Your Car-Horn. It comes with no coupons and has no hidden compartment for the storage of contraband nestled within its pages. Some things can be neither photographed nor foretold, and the B of the A-B trajectory starting with the Blueprints (if we assume that A stands for B, meaning *Blueprints for A Sparkling Tomorrow* and B stands for C, or *Consequences of Blueprints for a Sparkling Tomorrow*, that is) remains mysterious. Yet in spite of this somewhat knotty situation, we can offer a promise in the following form: *The Intangible Benefits of reading are sure to be grand, even if our Tomorrows fail to sparkle.*

THEMES OF THIS WORK

Is not love more potent than fate?

> *DUISBURG, Germany (AP) -- Crowds of people streaming into a techno music festival surged through an already jammed entry tunnel, setting off a panic that killed 18 people and injured 80 at an event meant to celebrate love and peace. The circumstances of the stampede Saturday at the famed Love Parade festival in Duisburg in western Germany were still not clear even hours after the chaos, but it appeared that some or most of the 18 had been crushed to death.*

Fate may send you swirling down a few unknown pathways, but love will crush 18 revelers to death in a tunnel. Yet while "Duisburg Discotheque" has lately become pejorative slang for any particularly self-destructive marriage, we ourselves maintain hope that love can yet be salvaged. After all, if it were not for love, what would be the fate of candy hearts? To put it another way, should love disappear, how would the other forms of madness ever stand a chance?

This work, then, is an attempt to systematically repair love, to channel its fearsome powers into Economic Progress rather than the senseless murder of harmless techno-obsessed tunnelgoers. We recognize the arduousness of this self-imposed assignment, but do not shy from it (though we must warn that by the volume's end we may have given up on it entirely, and instead begun an exhaustive monologue about the folly of contemporary transit).

There is love to be found down every avenue, in each crevice and crevasse alike. But perhaps love is, as was foretold, little more than "pissing with a hat on." If life is execrable or excretable, what can love possibly do to improve it? As Baron Rogers-Nimni, Marquess of Hampshire, was known to say loudly at parties, "one cannot long polish a turd." But one could outfit it with emergency brakes, *if one was willing to get one's hands dirty.*

All of this is inconsequential, however, for no question can be answered without first dealing with a series of additional questions. In order to find love, we must first survey human achievement. This work begins in the Cambrian, then, and ends in the Tunnel of Love.

The contents are (is) divided into four competing sections:

ELEGIES, *in which we mourn passed and passing phenomena;*
INCOMPOSSIBILITIES, *in which we establish the tensions of being;*
ORTHODOXIES, *in which we save received wisdom from itself;*
GROWTHS, *in which we build and play new social organs.*

These sections parallel the divine time-measurements of past, present (a), present (b), and future. The work is built upon a number of key artistic/temporal premises:

1. Certain things that have happened have been excellent. Others have not, and will not be discussed.

2. Certain things that currently happen are excellent. Others are not, but will be discussed.

3. Our future may contain many excellent things, but these must be noted to be realized.

We are not, then, the screaming red leftists that some are (and that some have accused us of becoming). The presence of the word "communitopia" in this work's abandoned former title should not suggest that we have any sympathy with the activities of communards or dystopians. Our careful studies of history have led to the distinct conclusion that brothers Marshall and Joe Stalin were among this century's most unpleasant autocrats. Never would we suggest the throwing of a bomb or the poisoning of a constable. Violence begets violence just as surely as Stanley begat Africa.

Much is worth preserving in our present system. The arts and sciences may generally be "harems of the useless," but that does not mean we would not venture back into our burning homes to save Dr. Spock's child-rearing manuals (or, for that matter, our first edition copy of William F. Buckley's *Getting It Right: Clever Phrases to Use When*

Engaging the Services of a Prostitute) from the conflagration.

But one need not accept the trough with the hay, so to speak. It is perfectly (indeed desirably) possible to both enjoy certain things and dislike others. One can be both nostalgic and progressive, malcontent and *bon-vivant*, libertarian and socialist. Dialectical history is written dialectically, and there is no shame in rejecting both Thing Number A and Thing Number B in favour of a hastily cobbled-together Thing Number C that synthesizes the useful elements in both A and B. Who could seriously think otherwise?

Yet it so often *is* thought otherwise. As I wander through the planets, I (the human) am confronted incessantly with demands to adopt a rigid and stable identity. Am I worker or Tory? Mod or rocker? Fisherman or Taxidermist?[3] Do I live in Japan or Macedonia? The answer, to all of these questions, is *both*. Extremes can be synthesized into little portable golden means.

None of this is to imply that we are not extremists. A sudden blast of cold extremism can be a brisk way to commence the weekday. When we go out plucking truths from the various tradition-trees, we need not only choose the most watery ones (indeed, seasoned truth-pluckers would be aghast.) No, one must harvest plump, fertile, and sumptuous truths, if one is to fathom a universe using them.

There is, then, no need for the reader to fear the possibility of moderation. We will be both reasonable and ghastly in alternating portions. Sometimes we will be both simultaneously, or Nimni will be one and Robinson the other. Either way, *do not question our bona fides unless you have a note from a superior.*

With that, please enjoy this lovingly-baked wisdom pie.

A Word on Guarantees

We will always tell it to you straight. This work is little more than a paper-flattened incarnation of the old-timey "Straight Talk Express," which keen Western history-buffs will recall was the first vessel to carry an information-cube more than a few hundred miles. No, we will speak neither in the letterer's traditional euphemism ("What a cunning hat" and similar obfuscatory substitutions, etc.) nor his devious cousin, the stout dysphemism. This may sound like ripe talk coming

3   See our discussion of "The Fisherman/Taxidermist," in Part IV, p. 175.

from two scholars traditionally noted for not only their hearty sprinklings of dysphemisms in text, but in fact their more-than-occasional
use of the ever-feared *cacophemism*, but we assure you that this time we
are serious. **We are never frivolous with stakes this high.**

Certainly, we have made mistakes. Our ill-timed prophesy that
Ceausescu was nothing but a "trifle upon a piffle upon a stick" made
us a bit of an intellectual egg-faced laughingstock several decades
back, as the clever reader will no doubt remember. But organisms are
destined for adaptation, and we have since learned to couch our implausible overstatements in terminology both broad and contiguous
enough to provide us with a cozy velvety cloak of Plausible Deni-
ability should the journalists or interrotrons ever come knocking at
our cabin windows. An unbreakable promise, then: **No matter how
many of our assertions may be untrue, we guarantee that few of
them are disprovable.**

Too, we have moved from the calaboose to the vanguard of the
passenger rail-train of American Social Thought. Initially proud of
our reputations as "card-carrying members of the après-garde,"[4] we
soon realized that we were languishing in the Political Philosophy de-
partments of fifth-tier liberal arts colleges ("the asthmatic anus of
American public life," as George Will memorably christened them in
a moment of soon-regretted candor), and effected a shift forthwith.
Every action we thenceafter took was conducted with the utmost feck
and aplomb. **We have not forged lightly our reputations.**

We do not fail to keep in mind the ominous onus that Eric
Hobsbawm once owed to Hannah Arendt: insofar as the "social
prophets" presume to be doing social science without adopting the
sciences' self-doubt, they will have successful careers as public in-
tellectuals and will never be held accountable for their vacuous and
unfalsifiable statements. More verisimillious words seldom got spoke,
and we carry a nasty infection of Hobsbawmism with us always.

With such cautions having been caveated, let us proceed section-
ward to Part I, in which we mourn things lost and suggest which ones
oughta hastily reappear.

---

4   A sobriquet assigned us by a certain late novelist, whom we initially disdained for his
    wanton florid proxility, and with whom our stylistic disagreements later escalated almost
    to the verge of letters-to-the-editor. We will not name him here, but will remark that he
    happens to (non-coincidentally) share his initials with the code letters of a major Texan
    airport and is a notorious practitioner in the art of the discursive citation.

I. Elegies

Before continuing, please contemplate Matthew 5:22:
"Whosoever shall say 'Thou fool!' shall be in danger of hell fire."

The past is neither candy-coated nor aerodynamically streamlined. Certain events within it have been nothing short of discouraging. The barbarism of Man, and to a lesser extent, Woman, has bubbled and raged since our ancestors first crawled from their sea-birth-tubes onto the land-space. We do not, therefore, venerate Tradition for the sake of Tradition. Some traditions are frivolous and moronic (see the inexplicable American pastime of catfish noodling), and we have neither qualms nor quennies about the tearing-down of the occasional stately Tudor manor-house to build a neon-dappled parking facility. The future must be both embraced and given a tickle!

Yet it cannot be argued that every single thing that has occurred, from the first sunrise of our people to their present workaday confusion, has been negative. Life is a raging seesaw of virtues and horrors, and should a child ever violently be flung from said seesaw onto the unforgiving concrete below, well, such is the price of doing business. Let not the occasional mass infanticide spoil an otherwise memorable picnic.

It is difficult, however, in an era when value judgments have been consigned to the banana-cupboard, to sort the sand from the rhinestones. Even if each and every American immediately concedes that The Past is a bit of a mixture as far as quantities of good things versus quantities of horrible things are concerned, what is to be done after this admission has been admitted?

We, Oren Nimni and Nathan J. Robinson, are of the opinion that the most adequate aqueduct down which to flow is as follows: Enumerate the good things and then preserve them by force. What did they (the past-creatures) all get correct, before they were violently wiped away? How can we live as they did, without having to wear any of those bloody silly costumes?

This section is elelogical, meaning that it is "of or pregnant with elegies." It currencies not in items that are still are, but items that *once were*, and ought to be brought back through a swift course of both imagination and legislation.

The beady-eyed reader will spot all manner of referentials to items s/he has read about in s/her un/cl/aunt's copy of the Harvard Classics. With nearly 1000 years of human civilization having come before us, no sentence in this work can be said to have not been spoken before. There is nothing new, as the gentleman said, excepting the sun. We hope, then, that what we lack in originality we may make up for in clear-eyed common sense. There are small philanthropies to be found in synthesizing the ancient wisdom, repackaging and monetizing it. We may have less to say than Plato or Postocles, but we have done it in a handsome hardback available commercially, which is more than can be said for old Posto and his pupil.

This section, like most others, has not been designed with the linear reader in mind. We consider this a book one more likely reads while taking a satisfying dump, or sneaking some light knowledge-acquisition during the annual on-site safety training review lecture while the Chief Inspector has turned from the podium to face his worthless charts. So take it leisurely, and without forward momentum. Dip in and paddle about! Don't ever change! Go and have a smoke!

A. Structure and Infrastructure

Attractive Things Well-Built

Today, most of the things that there are, are a blight. Take a step outside. Look down at your walkway. It is made of bricks. But does it *tell* you anything? Has it been designed with thought, with love, with edification of the bearer in mind? It hasn't. Most likely, it has been flung together with disdain. The workmen who built it hated themselves, and more importantly, hated you. As they arranged the paving-stones into makeshift floral seleidoscopes, do you know what they were thinking? They were thinking about tea, about their children, about their salaries, about the small tear in the underskirting they needed to mend upon returning "home." They were not thinking about how *your* walkway would help you find God.

Now take a step further and wander down it, tutting slowly. Look in your neighbors' mailboxes. Are they flimsy? They are. Do they sing? Do they escalate beyond mere "function"? Do they encourage the postman to insert letters with an ecstatic *joie-de-vivre*, or do they further wear him down and aggravate his risk of postal trauma?

If you continue your thing-stroll through the object-forest, you will find that much of what our blogs refer to as the oasis of civilization is actually caked in ugliness, turpentined in a viscous goo of thoughtlessness and ill-design. The objects and places human beings have decided to locate themselves within (and on) are unfit even for death or retirement.

Even the once-stately pancake house is today shoddily designed.

The International House of Pancakes should serve as a new United Nations. Yet when one enters its main hall, does one think to oneself "Ah, truly this is a temple worthy of the almighty flapjack?" No. One's happy little homunculus is rather more likely to emote the phrase "This floor appears to have something foul stuck to it."

Here is the point: if a breakfast house is not a cathedral, then what hope does the cathedral itself have now that God has turned out to be so thoroughly (and rightly) dead? If one's new religion is breakfast, yet breakfast-architecture debases the soul (pours syrup in one's soul-hole),[5] *what then*? By all means, replace the gods with syrups, let Aunt Jemima be our Saint Peter. But to correspondingly allow structure to become flabby, to stucco our buttresses, this truly tosses the pig out with the blanket. Even your humble authors, who have never ourselves tasted a pancake, recognize the height of the stakes (or the heat of the steaks, to elongate a breakfast metaphor).

But breakfast was not always so shabby. Once upon a time, they did indeed build diners of beauty; the Parthenon's bacon skillets were notoriously dripping in ancient grease. Today, it is impossible to imagine undertaking the construction of another Belgium, a country founded entirely for the purpose of serving waffles. How long has it been since America attempted to set a new record for the World's Largest Pancake, or to send it into space? A lament, then, for a time when we were willing to use a country's entire marble supply purely for the building of a city-sized cronut bakery.

"Do you dare to tell me how to build my own crematorium?" the impudent reader asks of us. Well, *yes*, we do. We believe in an objective architectural science; we adopt the position that there are "better" and "worse" crematoria, diners, and dinotoria (dinosaur crematoria). Just as "Rugby" is an objectively worse name to give one's daughter than "Lavender," there are buildings which fittingly pay tribute to God's glory, and there are those which defecate in His cabbage-patch.

Things must be built *well*, then, if architecture and suicide are to be different enterprises. But, as we shall explain carefully and with modest quantities of references to our own previously-published works, beauty is not enough. There must also be enormous pyramids.

---

5   Disclosure: the Soul Hole was briefly the name of a nightclub your authors were co-investors in with a certain well-known Republican blues guitarist.

PYRAMIDS, ETC.

There was a time when they built pyramids. *This is not that time.*

Intuitively, the above micro-statement is accepted by the public as largely correct (perhaps with minor exceptions; contrarians will always be tempted to shout "Aha, but I have seen a contemporary Louvre or two in my time!") But this is not the interesting question. The interesting question is: *why?* Why pyramids then and not pyramids now? Have we forgotten the triangle?

We have not. Triangles continue to appear, say goodbye, and reappear. They, in fact, do so at their leisure. Residents of America's Boston, loomed over as they are by a perpetual giant triangle, will have no illusion as to whether triangles have become as "outmoded" as certain trendular fashion correspondents have laughably alleged.

The problem, then, is not that the triangle has ceased to be valid, but that our triangles are simply too small. "If there are triangles," inquires the child, "what keeps us from having pyramids?" An astute question, but deluded. For what keeps us from having pyramids is the very ideological presupposition embedded in the child's words. Nothing keeps us, dear child! Stop presuming we are kept!

Could we build a pyramid if we got up and tried? We could. And yet, we could not, for so firm is our conviction that this is impossible that it thereby becomes impossible. Another prophecy eternally fulfilling and refulfilling itself, like a self-fellating duck sculpture.

But nevermind the wherefore, what of the why? Is a civilization improved by the addition of pyramids? We have never been Egyptologists, but we hazard that the answer is "oui."[6] The geodesic dome, publicly beloved as it may be, is not the only permissible shape. The construction of a pyramid or three (no pun intended) would certainly be less out-of-place than, say, the log flume in the Library of Congress or the White House's bowling alley. Were these additions acceptable? No. But the taxpayer satisfied himself with a tiny grumble and footed the expense nonetheless. Would he not do the same for a pyramid? Who, in truth, could say he would not? Or at the very least, ought not.

6 Trans.: "Yes."

GRAND ILLUSIONS

The pyramid, if built properly, is a very big object. But it is only one of an entire genus of phenomena that share a similar quality: the grand illusion.

Consider the gods. They hulked and fumigated over the Ancients, with constant promises to immanentize the miraculous. When the miraculous failed to immanentize, the gods produced all sorts of convenient alibis and mishearings. "Produce a *monocle*, you say? Why, of course!" Zeus would announce to the impatient worshiper, furnishing unnecessary eyewear. (In saying this, we do not mean to suggest that Zeus was a fraud, we are merely implying that there may be a reason Greece is today known mostly for feta and bankruptcy.)

But the fraudulence of gods does not discredit them. We would be seen as hypocritical were we to denounce fraud, given that a small sentence for such behavior is technically present on both of our resumes.[7] It is not that the gods misled their worshipers, for such behavior can simply be seen as charming and mischievous. Who would not a rakish god prefer? Not we. Rather, the problem is that by being very large, these gods ensured that *there could be no other large things.* For for a thing to be god-sized was a presumptuous slight upon the gods; thus did our chocolatiers enact a scale ranging from fun-sized to king-sized, with nothing beyond.

But "king-sized," is this best we can do? Many kings are very small. If we are to satisfy ourselves with never producing anything larger, we foreshorten our aspirations pitifully. It is the *grand* illusion that is truly grand. That fact was obvious to the 19th Century Human; today it seems almost madness. It would once not have been out of place to say that a cathedral is just as worthy an investment in the public health as a cutting-edge teaching hospital. How could a statement once seem so reasonable yet suddenly become so not? It is, we posit forcefully, the disappearance of the grand illusion.

---

[7]   We still maintain strongly that impersonating a racehorse for research purposes does not fall within the scope of the law's intent.

TAILFINS AND MICROBEAUTY

Each time we witness a contemporary car accident, we are struck by the tragic disappearance of tail fins as a feature of the American automobile. Once upon a time, when a man rolled off the lot in a fresh new Cadellarium or Chevrodeuce, he could expect to be trailing a fin the length of a five-iron. The fin's grace gave meaning to the driver's day and poise to his bearing. "Truly I am well-equipped for any emergency that requires the use of a tail-fin," he could have thought to himself.

Alas, such no longer remains much of the case. All it took was one impertinent passing child to ask "Sir, but are not tailfins an unnecessary luxury?" and the most elegant phenomenon in American motoring was cast to the wayside like an unwanted sibling. The virus of efficiency worms its way deeper into the American pineal gland.

No wonder one so rarely finds a pieman or a haberdasher these days. No wonder they demolished the velodrome and replaced our parents with non-committal warehouses. No wonder, if the tailfin went, so did the national parks soon follow. If we cannot value beauty above safety on the *micro* level, then how shall we fare with its promiscuous macro cousin?

This is the point in a gameshell. It is the microbeauties that make everyday life a treat. The tailfins that transfix us, the neon signs that buzz us into a warm fetal stupor. Small is not necessarily beautiful (consider the ant), but small beautiful things are that which makes life a tolerable pastime. Those who subscribe to sleek modernist styles and wear only one shirt their whole lives miss the entire point of existence; minimalism has no place in a very large universe. To drive an unornamented car is to get to one's destination without ever having really driven there, and we can think of nothing more tragic than inhabiting an improperly augmented state of being.

B. Society, Culture, & Animals

LACK OF HORSES

But even beauty has its opposite. Today we find ourselves amid a cataclysmic global hideousness, in which everyday objects repeatedly disprove the old aesthetic truism that *there are some shapes so ugly that they could not exist.* In fact, there is one shape whose ugliness is surpassed only by the very quantities of its existence. We refer (of course) to the horse.

It is clear from a glance at the horse that he is an unintended specimen. What gravy-brained gods could have dreamed up such a defective contraption and let it run loose? Not any worth inviting to the biannual church picnic. With clompers on its bottom, and gnashers lining its mouth, the horse gives us every reason to conclude that whatever deities may be found in whatever forgotten carrion-cupboards of the universe, they certainly do not exert much of their omnipotence on the provision of sensibly-designed fauna.

And yet: was ever thus the horse? Neigh. In fact, a time can be speculated upon *before even horses plagued mankind.* This placid pre-horse antiquity is seldom considered (for obvious yet non-obvious reasons), but that does not mean that it never once was. The horse has been around for so long that it becomes difficult to ponder a horseless paradise. Yet in precontemporaneous times, the very mention of a horse would have elicited silence from the creatures to which one spoke.

A lament, then, for those times when a man could set foot on a boulevard without worrying about manure, or wander a horse-ranch without seeing a single horse. For those times when one's shoemaker would not politely refuse a commission on the grounds that he was busy hammering out steel Us for the next half-a-year, and for those times when politicians with unusually lengthy faces needed not fear unkind equine comparisons from devious editorial cartoonists.

The horse cannot be executed, naturally. But there are more solutions to a problem than simply executing all parties involved. Exile, for instance, has proven an effective substitute for death.[8] While there are certainly egregious historical cases of unjust banishment,[9] there is no satisfactory proof that the horse could not be removed from society and sent someplace.

A Hearty *Weltanschauung*

It is easy, however, to lapse into simple laments for bygone absences of certain noxious animals. But it is not the case that if we were to draw up a list of disfavored earthly creatures, gather them two by two, place them in a great ark, and set it alight, we would thereby bring about a transcendent perfection. One cannot simply *do* one's way to the ideal; one must also correct one's *view*.

One of the primary distinctions between the past and our now is the sudden disappearance of the All-Consuming Explanatory Framework in favor of the Terrifying Set of Unanswerable Questions. The provision of worldviews was the foremost function of the religio-monarchical state. The point of a king is not to be useful, but to tell each of us what to think on matters from the spiritual to the geological. Today, with self-sovereignty having been put in charge, each of us must decide for ourselves what shape we think the Earth is, but at one time the ruling elites simply insisted that it was round *and that was that*. Yes, it's true that there is something fishy in being told what to think, and being released of one's obligation to share the monarch's obsessive affinity for tiny porcelain bulldogs is at least some small kind of relief. But we cannot fully share a cab with those who think the vanishing of all values into the cosmic void is a refreshing moral exhale.

8 *"Better dead than exported."* - Aphorism #1. Throughout this text we have littered a series of useful epigrams of our own devising. They are wise and catchy sayings, much like an advertising jingle or the fortune inside a cookie. See Appendix A for advice on their use. We hope these humble aphorisms might someday contribute to a new body of folk wisdom.

9 We refer especially to cases in which simple forthright language at a proseminar is inexplicably deemed adequate grounds for the imposition of an involuntary sabbatical.

Think of how Napoleon solved a problem, whether it be a difficult set of sums or a debate over where to carefully place an unattractive tapestry so as not to offend the weaver while still keeping it from the eyes of guests. Napoleon would have only one question to pose himself: "What would the most French answer be?" And whatsoever it was, thatsoever he would do.

But today, we are left with nothing to guide us through our weekly identity crises and employment disputes save that wisdom found in the new age mottos printed on condiment packets and in the Horoscopes page of the *New Orleans Review of Books*. It is a paltry substitute indeed for a Bible or Fihirist.

Any child can open his school's tattered copy of the colonial classic *Worldviews Now and Then* (Yankee Dollar Pub. Co., 1950) and see what the world looked like through the eyes of dead humans. It was so easy to be a Jainist once! But one will immediately be struck, not just by the fact that all of the gods have been slaughtered and piled into a mound, but by how shabby and unimpressive they now appear to have been to begin with.

The challenge, then, is not to resurrect the old worldviews, for Jesus lied when he defensively insisted one could be resurrected without thereby being classified a zombie. Zombie politics make rotten bedfellows. No, the challenge is to develop new and mightier *Weltanschauungs*, while nevertheless eulogizing the old and lamenting the untimely passing of their time.

THE EXPLORATORY IMPULSE

Too many times a day we are fed the question "But, Daddy, what is left for an explorer today, now that the ice shelves have melted/tumbled and outer space has been found to be a deathly bore?" A peculiar inquiry, little child, but not one that we are insensitive to.

It is true that meetings of the Geographic Society have lately lost some of their vim. In the last century, the committee might have been debating a resolution on how many boxes of wolfmeat and snowshoes would be needed by this or that neo-Shackleton in his quest to survey some new bog or delta. Today, the Society's fiercest arguments are taxing retreads of long-running schisms over which map projection gives Lady Earth her superior shapeliness.

A regrettable state of affairs, then. But not necessarily one that is to be regretted. For the exhaustion of exploration is far more a matter of method than location. The folly has been to focus on *where* to explore, rather than *how* to explore. Yes, it is true that all the old techniques have antiquated themselves. One can no longer just press gang a sherpa, pack a provision-basket, and drive a dogsled to go and find some new cannibals. Yet that has never been the real purpose of exploring, only its most visible symptom.

The real purpose of exploration is, has always been, will ever be, *self-orientation*. Like Lady Godiva's windsock, the explorer knows which way to point. We don't know where we are until we have a look around. We can't decide to go and look off that promontory until we realize what a promontory is and whether it is, in fact, the one we are talking about.

Thus to give up on being explorers is to give up on self-knowledge; a rambling hike through the mental chaparral has never ill-served a truth-seeker. There are ways of exploring without leaving one's mother's futon, so to speak. For exploration to be a thing so dead, it is astonishingly alive. What we must explore are possibilities rather than places; we must posit a million futures and find the canals and passageways that lead to each. Where classical exploration is concerned, we can surrender the act without giving up the impulse. Accuse us of oversimplifying if you must, but every major theorist must have an explanatory mechanism, capable of putting enormously complicated phenomena into childishly basic terms.

COMMITMENT TO A THING

But we must retire from merely altering perspectives, and return to altering the question of *what one does once those perspectives have been altered*. Related to *seeing* things monolithically is taking monolithic *action*. The most important part of having a political belief is not what the belief itself is. Instead, what matters is that a person be unwavering in their commitment to the belief. It may have surprised much of the nation when Obama criminalized monogamy, but it was admirable as a first show of true principle.

Media celebrities have political commitments, too. Consider this passage from the autobiography of Wolf Blitzer:

> *Some days I would have given anything to have died in that tornado. But I realized that that wasn't my job. My job was to make the news exciting, regardless of consequence. I knew that my politics were the politics of charisma, that I finally had a purpose and here it was.*[10]

Blitzer reveals we never need ask the question "How can a journalist serve both his Ego and himself?" But the passage is quoted here primarily for its emphasis on commitment. To blitz oneself is to turn oneself over to the ecstatic life, to take on a supernaturally perfected human form.

An inevitable mistake will be made by the reader, who will surely assume that by speaking of the "committed man" we refer somehow to poets. Surely a poet, having turned his life over to the production of the erratic phrase, and having eliminated so completely his economic usefulness, is the exemplar of the committ-ee.

To the contrary, however, our views on poetry border on the downright Platonic. Any reader of our previous works will already know our feelings on that bearded poet and confidence man[11] Ginsberg, a man who had the impudence to assert that the great minds of his generation were to be found among its Columbia undergraduate English majors.

No, by Commitment to a Thing we refer to a different kind of committed human altogether, what refer to as the One Who Gets Things Done. For example, we knew the first time we drank the contents of a lava lamp that we had made a bad decision, but **we stood by it**.

Consider our friend Malcolm X. His irresistible looks, his eloquence, his defiant and uncompromising rejection of the dominant culture; all combine to create the most fearsome of revolutionary

---

10   from Wolf Blitzer, *Hungry for the Wolf: My 30 Years 'Blitzing' Radio, TV, and Print Journalism* (CNN In-House Press, 2008), p. 430.

11   Is every poet a confidence man? We dare not hazard an answer here, for fear of having our work reviewed by poets, but interested readers are encouraged to peruse our recent academic article "All Poets Are Confidence Men: Why Every Poet is a Confidence Man," *Quarterly Journal of the Society for the Deportation of Poets*, Vol. 8, Iss. 2. (May 2014), pp.109-114.

figures. Malcolm is humanity at its most optimized, and to revere him is to revere ourselves selflessly. The honest aspirant therefore asks himself the question: "Who is your own Malcolm X?" Having an answer may prove valuable.

Yet it need hardly be pointed out that Malcolm X is dead. One cannot go and shake his hand, lest one is willing to provoke the ire of the carceral state.[12] The creation of substitute Malcolms is therefore a moral imperative. We eulogize the intensity of his commitment, but the real task is to redevelop that self-same commitment to things of our very own. Until then, in some sense, we re-assassinate Malcolm X every day through our neglect of him.

But having examined the desirability of and intensity of commitment to things, let us look at what some of the things themselves might be. The past contained not just attitudes, but tendencies.

Uniforms Without Uniformity

If there is one thing we have learned from historical example, it is that the good done by enforced uniformity cannot be overstated. It cannot, in fact, even be stated. For the benefits of uniformity do not occur in words, but in acts, and acts cannot be placed upon a printed page.

Here is what we mean to say: the popular consensus on uniforms is that they are and always have been a mistake. "This," says the conductor, "is the Age of Individuality, in which we shun that which is grey and consistent in favor of the wildly varying and kaleidoscopic." Each person is supposed to go and find them or theirself, and then decide what he or she would like to do today. *Each is a self-actualized version of herself* rather than a malleable sprog in a vast cosmic sphygmomanometer.

But is it desirable for each of us to be our own person? Does that not create a very large number of people? It is understandable that after the 20th Century horrors wrought by the Maoist enforcement of denim, we would seek refuge in the sartorially diverse. But is individualism the remedy, or does it lead to twice-a-day identity crises in everyone from stockbrokers to wharfmongers?

12   See N.Y. PENAL LAW § 145.22, "Cemetery Desecration in the 2nd Degree."

We posit the latter. For we have always responded to the question "Who would you like to be today?" with a resounding "Whoever walks through the door first, thank you." Some of us *do not wish to have to self-determine*. Not that we wish to be handed tickets to a preassigned concert of the self from a centralized government identity-dispensing booth. But there are still certain shows we would like to go to without having to build the entire stadium ourselves.

To gild the tips of the point: it is not that the state should impenitentiarate whosoever its citizens dare not to wear the national uniform. But we do believe the virtues of mandatory stylistic conformity have been overlooked in our understandable focus on its considerable vices. For would not a streetcorner look a dash more coordinated and enlivening if all of its denizens sported a standardized lavender tunic? It is difficult to argue that it would not.

Do not execute dissenters, then, but do consider perhaps asking your neighbor to coordinate outfits with you tomorrow.

Gentlemanism and Flâneurship

We cannot think about *whether* to wear what, however, without thinking about how to consider *what* to wear. Yes, we should all wear uniforms, fine be it. But in theory a uniform can range from the minimalist fig leaf to the maximalist solid-gold spacesuit. We can agree we must wear the same things, but which of the same things ought we wear?

The inquiry dovetails neatly with another pet fixation of ours: the theory of gentlemanship. For a gentleman knows ever and always what to wear. He never wears a striped tie before Armistice Day. He never places a carnation in his buttonhole when what is wanted is a chrysanthemum. He would not take his top hat off in a mausoleum. The gentleman, in short, is a specimen of utmost grace and decorum in matters of the cloth.

But where is the gentleman today? Certainly not in the city square, for when one stops by one is struck not only by the positive presence of pigeons, but by the positive absence of gentlemen. A fop in a cravat is these days as elusive a specimen as a good shark fin soup. These days one can even attend a country wedding without ever being kissed on the forefinger by a minor earl.

Can the situation persist? It can and it shall. But those of us who are powerless to stop it may take comfort, for we are not powerless to stop it. For gentlemanship is a choice, and thus it can be chosen. Each day we go out, we either enact or fail to enact our position as a gentleman. If I decide to go out with chaps where my lapels should be, I have made a *conscious contrivance* to refuse the gentlemanly virtues. If, on the other hand, I place a handkerchief in my brassiere pocket, and carefully sharpen my poking-cane, I may yet bring a sense of sensibility back to downtown Sarasota (or whichever other city I may happen to consider myself to be from).[13]

So, too, the flâneur, the idle aesthete who loafs in cafés. Where has his unique variety gone? The answer, tragically, is that like so many of us, he has been forced to seek employment. But he who says "Be pragmatic" is in reality chanting "Surrender!" and the raw economic facts of the day should never stand between a society and its dreams of loafing. The gentleman and the flâneur must be preserved *no matter what the cost in material resources or human lives*, for there are some values so worth saving that all other values ought gladly to be exiled in order to ensure their eternal persistence.

He who is not a gentleman is neither gentle nor a man. But the true Good News is that each of us has within herself the inner capacity for gentlemanship, with the only obstacle to its enactment being the vast time consuming tedium of teaching oneself which shade of beige garter best complements a summer sweater.

THE FORCIBLE IMPOSITION OF VALUES ON ONE ANOTHER

Like all other rural youths, our childhoods felt dominated by two ever-present truths: never criticize the President and always stay true to the Timeless Necessities. Sadly our adult lives have seen most of the necessities become unnecessary, and we have come to realize that occasionally the President behaves like a little boy in a barnyard, kicking pigs for fun.

We can't do much for the President. But we can do something for timeless values. The problem, as we see it, has been an abundance of

---

13 Yet in case the reader has become alarmed by this portion of the text, note that we are no dogmatic sartorialists. As proof, one need only spend some time with Aphorism #2: *"Daylight comes through even the most misshapen buttonhole."*

respect. Today, it is seen as fashionable to praise any old papier-mâché puppet as a creative triumph, and any old noise as a symphony. But some things are better than others. Our values must be enforced.

"Aha!" says the nitpicking picnicker; "There I've got you! For your values are but arrogant personal insistences! They have no transcendent core!" But he has not got us at all. For we anticipated this objection and have already incorporated it into our position.

In fact, any objection that could conceivably be made has already been foreseen and forestalled. Those who wish to challenge us will find themselves like the man in the story who attempts to tame the wave.

Of course our moral values are utterly nonrational; they are withdrawn slowly from the anus like a magician's handkerchief. But why should we go disbelieving a thing simply because its origins be sphincteral or magical? Our relativist neighbours make the same colossal reasoning error in every cease-and-desist notice they slip under our office door. "The source of moral truth is not Ultimate," they say, "therefore we must respect various claims to moral truth." But this is the most violent non sequitur since the Jacobins decided that widespread wanton beheading was the only logical reply to an oblivious remark about pastry. For the claim that we ought to respect people itself appeals to an Ultimate; what persuasive reason can you give me, if my happy little subjective code dictates that I ought not respect people, for me to respect people by following some universalized code of respectful subjectivities?

Thus, because we Nimnis/Robinsons believe others' values and aspirations are much worse than our own, we resolve to use all of our academic energies toward their destruction. This is not to say that we deny others their current *right* to philosophies different from our own, merely to say that we find those philosophies to be suicidally mistaken and in need of ruthless assault.

We are therefore unashamed dogmatists, justified by both reason and common sense. And yet somehow, those who are dogmatic have historically also been obnoxious. If we look at how prescriptivists have historically offered reading lists for college classrooms, we see that they are not only firmly convinced of their rightness, but also insist on being incorrigible pricks about it.

Thus, we would like to turn now to the example of the Great Books of the Western Canon question, in order to see how one might forge a compromise, a prescriptivism without prickliness, a universal value system behind which everyone can joyfully get.

Rather Good Books

The drawbacks of a Great Books education are well-known, but the common error has been to conclude that these necessitate the abandonment of literacy altogether. If it is foolish to foist Great Books on children, what use are books?

A sensible question, of course, and one that is difficult to answer. We were never on board with the idea of promoting Herodotus to the under-sevens. To University of Chicago professor Allan Bloom's death, we said "Ah, not a moment too soon!" "Away with theorists" has always been our position, save for one or two obvious exceptions.

Yet if we are about one thing, it is solutions. And the solution to books is not fewer books. No, the real inquiry has never been "How many books should be allowed?" which is a more irrational question than it may initially seem. The real question is "What is an education for?" And in asking the latter, we may reorient the Great Books question not to be about certain *volumes* of book, but certain *types* of volumes of book.

We may turn to an example, so that the mousse of abstraction congeals into the cobbler of concreteness. *Timon of Athens* is a Great play. The *Necronomicon* is a Great book. *Bell Tether* is a Great film. The Undertones are a Great band. But how many pillars does it take to create a pantheon? Professors who should know better think making a canon is as simple a business as making a cannon. Even junior apprentice cannonsmiths would scoff mightily!

The cannon/canon disjuncture isn't nearly so simple as the academics presume. One has an additional n, and as each of us was forced to remember as a young pupil, n does not equal zero. It's not nothing, to put it differently. Canon formation is like the multi-decade elaboration of a coral reef; *we do not know what is in it until we finally snorkel the reef.*

Instead of the senseless prescriptivism of a Great Books program, then, we recommend a more modest alternative: some Rather

Good books. Requirements melt into recommendations, laws into gentle prods. Don't tell the children that Socrates was a good man; he wasn't. Don't tell them that if it's not in the original Latin, it's not worth the time of the day of the paper it's printed on. Don't hand them a list and say "Here is a list"; hide the list and tell them to go and find it. When they return, list in hand, tell them that that is the wrong list, and send them back into the thicket to quest for another. "The real list," you will tell them as they emerge empty-handed and bramble-thistled after a days-long trawl through the underbrush, "is invisible. It is in your heart." And lo, though they may curse your very name, ultimately they will be forced to agree that these are indeed some Rather Good books.

C. Political Arrangements

EUROPE

It is ordinarily considered indecorous to beat one's reader bluntly, but there is a frank query we must presently drop with an echoing thud: *Is there anything more European than Europe?* Addressing one's public with an inquiry so abrupt is to be apologized for, but since we reason that most will be reading this book voluntarily, they are not without recourse if they find themselves revolted.

But the European question is an almighty one, a problem the size of a continent. At the root of present American political anguish is the identity question: How European do we truly wish to be? We face the glaring tension between our beloved inheritance of Shakespeare and royal babies and our firm insistence as a nation upon the cultural necessity of pitilessly abandoning the elderly and impoverished. It seems as if one cannot enjoy Oktoberfest or the films of Gérard Depardieu without automatically signing oneself up for a generous welfare state.

How then, can we be *just as European as necessary* and no more? That is the timeless American anxiety, the one over which centuries of blood and newsprint have been poured. Neither a European nor a peon do we wish to be.

But if we take a step back from the plateau of nationalistic intoxication, we see that the question is a dimwitted one. Why would one strive to be consciously *less* European, when we have not yet even cobblestoned our streets or mastered the art of the pastry? Nobody is asking America to wear the Kaiser's silly helmet or sit through the music of Wagner; they merely wish to have our President rechristened the Prime Minister, a request easily satisfied.

No, it is Europe for which we should be nostalgic and not the other way around. Each U.S. state should be renamed after a European nation, each television station should conduct its broadcasts only in minor Slavic languages. If the grave lessons of Nazism are truly to be absorbed, it is not merely necessary to appreciate or have affection for Europe, but to veritably become it. Anything less is a disservice to Hitler's millions of victims.

A FASCISM OF NEIGHBOURS

It is impossible to speak of Europe without speaking of fasces. But it is impossible to escape the fact that fascism's reputation has tended toward the negative. Yet is it not nevertheless true that fascism, at its root, connotes simple neighborliness? I am a fascist, you are a fascist, therefore we together are fasc*ism*, and all the world is better off for it being so.

Do not muddy our gist. Nobody likes the firm suede boot of state power doing aggressive pirouettes upon his/her neckflesh. Yet we insist that there nevertheless remains a small Colonel of Truth among the ranks of fascist regimentist order. It is our job to find that Colonel and make him talk.

None of this is to defend fascism *qua* itself. Nobody is less unaware than we of the peril faced by our own kind of truth-teller under a fascist experiment. First they come for the philosophers, for the cunning politician knows that nothing is more influential than philosophy during a time of crisis. When Mussolini comes to Washington, do we think he will give the time of day to two wise and thoughtful blueprinters? We are under no such delusion.

We know fascism has not just a negative consequence hither and thither, but is a veritable mudslide of unreasonableness sweeping away the tiny ramshackle hut of humanitarian universalism. Nobody wishes to see such a hut destroyed. Who are we to say to the man roughly interrogated by the regime's soldiers that in his fuming antipathy toward the present condition he should nevertheless note how well-polished the torturer's coat-buttons were?

Yet it would be wrong to say that fascists were all unremitting brutes, when some of them liked the opera. No absolute statement holds absolutely, and to replace the absolutism of the fascist state with the absolutism of an antifascist nonstate is to maintain that very absolute that so soured us on the fascists to begin with. What was fascism *about*, really? It was about the undoing of variety. Thus complete consensus against fascism perpetuates *that very same stifling of the multifarious*. In telling the death squads to go home and reassess their careers, we should make sure they do not return the next morning having resolved to become truth squads. All fascists are brown-nosers; the meat of the task is not to alter the colour of nose, but to blow it.

Ditch the fascism, then, but keep the neighbors. We can still be nice to our friends, even if we are not members of the same violent paramilitary group.

NEIGHBORLINESS

Many times have we been surprised from behind by the kindnesses of a *clean and pleasant stranger*. Those that live around us, that form the crusts on the loaf of our existence (insofar as they are probably good for us, but are generally irritating and inconvenient to get rid of with only one's butterknife), are a concept requiring considerable inquiry.

Our neighbors fascinate us, not only through binoculars at night, but at the community center and the fishin' hole during the day. All that we are, we are thanks to a loving towns-worth of Cub Scout Leaders and Country Preachers. These moral stallions taught us to be proud of ourselves, and to never be ashamed of the hideousness of our personalities. While the other children would giggle and sneer when we wore waistcoats to the beach, or expounded on the virtues of translucent housing during classtime show-and-tells, our neighbors were there to comfort us and reinforce our most cherished unfashionable beliefs. We shall never forget that it was old Mrs. Hašek, who when we attempted to turn the jungle gym into a geodesic greenhouse, placated the other children with her molasses cookies so as to keep the mob from plucking the eyes from our heads with pilfered kitchen-tongs. (It is no ill reflection on Mrs. Hašek, of course, that two days later when the children came again, she was at a meeting of the Firemen's Wives, and the cookieless monsters at last succeeded in

their wretched task; though their chants of "Out, vile jelly!" will res-
onate in our brainspaces till death, the warm goo of Hašek-molasses
will remain similarly upon our tongues.)

The case of Mrs. Hašek is no more than a mere anecdote, of
course, and we recognize that there is no Absolute Guarantor that a
neighbor will actually be a charming bloke, and will not take a hack-
saw to your prized Siamese Hedge merely because it skirts the outer
limits of the property line. It is true that a neighbor can just as easily
be an arsonist as a fireman. Yet it is also true that should an arson-
ist-neighbor be one's own neighbor, instead of someone else's, one is
far less likely to be on the receiving end of said arsonist's particular
arson-madness. (After all, the best arsonists set fires that spread, yet
what sensible arsonist would wish to see his arson-base destroyed?)

Neighborliness therefore acts as a buffer against the more savage and dismantling qualities of an Individualized society. We use our
neighbors as human shields against themselves. If, say, I am a murder-
er, and thus get my jollies from murdering people, and I proceed to
murder those that live in the houses opposite mine and on either side
of mine, who will look after my cat while I am fleeing the authorities?
Not my neighbors, not if I have gone ahead and corpsified them.
Community is necessary.

Of course, some of us are *not* murderers; we do not subscribe to
the "massacre your loved ones" set of behavioral incentives provided
us by Western Capitalism. What is in it for us? Ah, but you forget the
cat, don't you? We have cats just the same as killers do, and sometimes
we, too must flee our own homes. The world's first self-resolving paradox.

Utopians and Utopianists

GRENOBLE, France — A utopian dream of a new urban community, built here in the 1970s, had slowly degraded into a poor neighborhood plagued by aimless youths before it finally burst into flames three weeks ago.

Yes, all of our Utopias tend to wind up aflame. But our central holding is that this does not negate their core moral principleism, just as the constant perpetration of vast slaughters by democratic states does not erode the view that mass political participation is the enemy of vast slaughter.

Once upon a few yesterdays ago, Utopianism was a blossoming thistleseed, popular among workers and wastrels in equal heaps. Today, ask your average barge riveter or artichoke harvester the way to Utopia, and she will look at you as if you have just asked "Why is math?" That is to say, she will neither be enthused nor amused.

None of the contemporary political organizations (which call themselves "parties" even though they are in fact quite the opposite) offers a Utopian platform, or even a flimsy plasterboard Utopian plank in the party manifesto. The National Marine Life Union Party (NMLUP) did at its first convention in the early 1990's include a so-called "Utopian footnote" on a single page of its Booklet of Principles. But as anyone with rigorous academic training could have told these misguided marine life in a sea-second, a footnote is barely a footnote until it consumes at least ⅔ of a standard printed page.

"Utopia is not a destination, but a safari," the mystics repeat. But this is capital-h hooey. Everyone enjoys the occasional rhinoceros hunt, but that doesn't make collecting wild animal carcasses the equivalent of socialism. To suggest otherwise would be, we posit, unreasonable. For Utopia is indeed a destination, the fact is in its very name. If utopia was no place at all, where would we be left?

The Democratic Party has failed us. It emits the oozy syrup of defeat like a sculpture of a micturating cherub. It offers no attractive proposals, thus it is little wonder that the attractive do not propose to one another at its fundraisers. A scad of leaflets and diagnostics has issued forth recently on the subject of Where The Democratic Party Has Failed And Why. But each of these is at least partial guano, because each fails to recognize the rôle of Utopia-denial in the party's downfall and shrivelment.

It is crucial that a ship know where it is going, for otherwise it will likely not get there. But ask even a sensible Democrat "What is your party's greatest and most fervent dream?" and you will get a milky stare. *They do not know the answer to the very question they are supposed to be answering.* "To what end?" is the central inquiry of a reflective political life, yet those who live politically have little to say on the subject of interstellar transit policy in the 24th Century, or how architecture should look once poverty is eliminated. Should humans live in giant beehives? The questions that must be asked are instead being left to artists. And artists, it need not be noted, have never legislated a spending amendment or formalized an important arms treaty. In fact, they have done quite the opposite.

Thus: we cannot leave Utopia to the Utopians, for they are dreamers and eccentrics. To be a Utopianist must no longer be a capital crime, but a very precondition of one's emergence as a political actor. Without a Utopia toward which to row, the canoe of state will flounder and capsize before the coxswain completes his first shanty.

INFINITE POSSIBILITY

It is ironic that today we find it infinitely impossible to imagine a time in which the impossible seemed possible to imagine infinitely. Yet this is indeed our situation. We do not even know what it may have felt like to conceive of that of which we cannot conceive. The depth of this tragedy cannot be overmeasured; it means inhabiting a Myopia instead of a Utopia.

Poverty of imagination is not proof of impossibility. That one cannot conceive of a thing does not make it an impossible thing, it simply means one is bereft of foresight. This is so tautological as to border on the self-evident, and yet serious philosophers[14] have somehow

---

14   We refer here to a gratuitously unpleasant "review" that appeared in the pages of De-ontology Today, in which a certain obscure British utilitarian (not a deontologist, mind you; the editor's brains had apparently taken their sabbatical during the production of Vol. 8 Issue 2) mustered the stones to label our nonfiction novella *Imaginatory Poverty: A Story in Two Hundred Syllogisms* (Demilune Press, $12.50, out of print [for copies, please contact authors; many available]) a "tedious exercise that reads like the abandoned Intro to Logic homework of a particularly dim and unpromising freshman." In reply, we would note only that this professor's belief that exercise is tedious is fully reflected in the state of his figure.

managed to find the gall to brand it unserious.[15]

When that small mimeographed journal *Proletarian Outlook* responded to the timeless left-inquiry "What is to be done?" with the answer "Nothing. Absolutely nothing," we first found ourselves refreshed by the unexpected shower of pessimistic piss. But life is not War, and we fear the editors may have simultaneously undersold and oversold themselves.

There is, in fact, much to be done. It is simply that most of it is impossible. But a *true* proletarian outlook does not allow that obnoxious truth to become intrusive. Yesterday's journal-editors would never have let fact come in the way of fancies. "Yes, yes, it is impossible," they would snarl at the disbelieving copy-boy. "What on earth does that have to do with the question of whether it must be done?" And the copy-boy, goggle-jawed and wordless, would be left to ponder this impenetrable non-paradox.

That is not what happens today. Today, when some imperious neophyte dares to ask the Chairman "but what about Major Obstacle X," the gutless Chairmen, himself indoctrinated in a similarly senseless pragmatism, will scratch his pate and mutter "Gee, I guess we hadn't thought of that." Then he will behave as if this somehow leads to the conclusion that the Party ought to come to its senses and change courses. The Party must never come to its senses; that is the whole point. *If one is not trying to possibilize the impossible, then what is one trying to do?*[16]

---

15 Lest we appear to be casting the consequentialist baby out with the utilitarian bathwater, we should note here that a certain respected mousy Australian philosopher has been consistently kind in providing notes on our excursions into ethical theory. Though we may differ with this gentleman on the subject of horses, after a string of dispiriting encounters with academic pettiness and snubbery, it was a welcome and encouraging change to receive his offer to provide written comments on our manuscript on the condition that we never email him again.

16 Aphorism #3.

Towards a Gradualist Formalism

Nothing good comes gradually. Progress is a sudden and merciless tsunami, that wells up and destroys the beachfront resort of backwardness before the sunbathers even have time to put on their fleeing-sandals. Long has it been our policy to scoff at those who suggest that sometimes things ought to be implemented piecemeal over a reasonable timeline. After all, was it not Martin Luther King himself who said "I want what I want, and I want it now"? There can be no room for ditherance where lives are on the line. Tarrying is for bishops.

And yet: parts of us have always been sluggish. Was it not us very Nimnis and Robinsons who, when asked by our former partners to consider rousing ourselves before noon on weekdays, replied with a strenuous insistence that sometimes things ought to be implemented piecemeal over a reasonable timeline? We will freely admit that it was, and that this tiny spoonful of massive hypocrisy requires accounting for in any serious philosophy of historical change.

How then, do we circle the square? On a tiny motorbike? No. Rather, we have developed an elaborate theory that allows us to insist there are no contradictions between our two seemingly contradictory positions. It is this: *the theory of gradualist formalism.* Yes, progress comes in big waves, but that does not mean all formal processes may be banished to the scullery like an insolent valet. Just as the fact that a party will someday end does not mean that one may come wearing leather chaps instead of tie and tails, the fact that progress generally bursts onto the scene uninvited does not mean one should not polish the china, if you get the cut of our drift.

The key term here is "process." Break the word into its component parts. Pro-*cess*. And yet to believe in the necessity of process does not entail the favoring of sewage. No, the point here is that *everything unfolds via process.* Process is a balm upon the severed lip of turmoil. One cannot control the turmoil, just as one cannot glue the lip back on, but one can apply a healing wax to the gushing wound. To give a more oblique metaphor: process is akin to a chilled beverage in a hot place. It is refreshing in the summer, but *it is just as cold during the bleak winters.*

How could one organize a society without a process? A thing must unfold, not unfurl. Change without process would be like a gymnasium without a spinal bar. *Not very useful at all*, at least if one's chiropractor has insisted that the sedentary docility of academic life has turned one's backbone into an unrecognizable tangled squiggle requiring an astonishingly expensive and time-consuming set of chiropractic remedies that one's spiteful spouse expresses a callously unsupportive skepticism towards.

Process, then, remains necessary. And through gradualizing our formalism, we develop a *practical* way of mounting the tsunami. We find our way to recognizing that simple truth: *everything is free so long as one is willing to work for it.*

AGAINST A FORMALIST GRADUALISM

Our notion of gradualist formalism, robust and incontrovertible as it may be, has come in for the occasional scabrous academic libel by certain discreditable Marxist geographers.[17] We have repeatedly suffered the tedious misfortune of sharing both discussion panels and awards-banquet-luncheon tables with one Prof. Harvey of New York's "city university."[18] The hosts of such events, oblivious to the even the most basic academic disagreements within post-Left utopian economics, mistakenly assume that we share Harvey's every illformed and dust-coated premise, or even his entire dessicated philosophical edifice. But *we could not disagree with him more mildly*. For we well remember when Harvey, in a naked attempt to flatter his flatterers,[19] insisted that in the future there would be no trains, and that *"locomotion must be slow."*[20]

---

17 Note: our text's original draft included sarcastic quotation marks ("geographers"), but certain capricious editors have a warped notion of the standards for professionalism and courtesy in academic writing.

18 An irrational conceit to begin with, for as we have demonstrated, the city itself is a university.

19 to pun more cunningly, to batter his betters

20 D. Harvey, *Seventeen Contradictions and the End of Capitalism* (Leftist Drivel Press: Brooklyn, 2014) We would note unkindly that the passage comes in Prof. Harvey's three-page section on "Solutions" after a two-hundred-and-fifty page book on "Problems." (*BfaST*, by contrast, is all solutions, with no problems.) This ungenerous observation would not have been necessary had the esteemed geographer taken in better humour our gentle dinner-table ribbing about the manner in which his beard caused him to resemble an emaciated communist Santa Claus.

Who could think it would be wise to place us on a panel with such a man? We do not value slowness for the sake of slowness, and even more we do not value M. Harvey's called for reified slowness. The world has been shown to be an undulating form, twisting and gyrating under our feet as the sand gently suckles the sky. Slow locomotion only leads to slow-commotion.

Conflating the gradually formalist with the formally gradualist would thus be the height of humiliating academic *faux pas*. Many a time have we had to use the opportunity of a speaker's Q&A session to correct this particular ignorant conflation. Inevitably, we are met with the wearying misguided reply, "Is that a question, sir?", from lecturers so witless that they do not even realize when they have been outwitted. Of course it isn't a question, it is a five-to-seven minute correction of a small tangential theoretical point. To ask whether our contribution is a question entirely misses the substance of the thing. We value inquiry for its own sake of course, but inquiry into inquiries is simply ludicrously excessive metatheorizing.

Let there be no confusion, then. A quotation may serve to buttress the point. Here is what Quincey had to say on the subject:

> To formalize oneself gradually is the task of social investigation. Statistics and photographs can provide us with strong correlative function, but they cannot guide us in the selection of ultimate value. Thus the choice of by what pace theory is to operate can only be a function of socially specific contexts, each contingent and self-reifying. (Quincey, 2010)

That, then, is the liqueur that Harvey and his "comrades" in the Marxian tradition spectacularly fail to slurp. Because they believe contradictions must be resolved, through social revolution or otherwise, they do not see how the only non-contradictory position possible might be the constant contradiction of oneself. Formalist gradualism, then, has seen its day and belongs in the relic cupboard with yesterday's popular golfing journals.

II. INCOMPOSSIBILITIES

INCOMPOSSIBLE, adj. Unable to exist if something else exists.
Two things are incompossible when the world of being has scope enough for one
of them, but not enough for both- as Walt Whitman's poetry and God's mercy
to man. Incompossibility, it will be seen, is only incompatibility let loose. Instead
of such low language as "Go heel yourself- I mean to kill you on sight," the
words, "Sir, we are incompossible," would convey an equally significant
intimation and in stately courtesy are altogether superior.
- Ambrose Bierce

"Ghosts should not rule and oppress this world,
which belongs only to the living."
- M. Bakunin

"In This Weather, Even the Melons Are in Peril"
- Headline, *The New York Times*, July 24th, 2010.

I f *even the melons* are in peril, what is to become of us, the lowly consumer or consumerette? Here in Part II, we examine that which can no longer be tolerated, those things that are *incompossible* with human living in the Bierciean sense.

Sometimes two forces are so oppositional that they become contrary, and sometimes so contrary that they become argumentative, and sometimes so argumentative that they simply *cannot* hold positions in the same academic department without foreclosing permanently the possibility of a non-highly-discomforting faculty meeting. These are the phenomena with which this chapter shall deal.

There are things in this world without the nonexistence of which we cannot survive. Most of these are safely buried in lagoons and trenches. But occasionally, if one does not keep one's eyes trimly peeled, they bubble up and get trapped in a ventricle. What happens then? What happens when something which *cannot exist* if humankind is to prosper, nevertheless exists? Why, it must be given an ice-axe to the temple (or be denounced in an editorial).

How much misery could have been avoided if certain undesirables had been snuffed as toddlers? If James Garfield had been convinced as a tyke that the presidency was an ignoble calling, our country might have been spared both one of its most traumatic assassinations and two animated-blockbuster-films-worth of the miserable antihumor of the man's feline namesake. If Mr. and Mrs. Žižek had never decided that they were right for one another, how many pages of neo-Hegelian film analysis might our poor civilization have been spared?

There is an obvious counterpoint. Yes, if you go around forestalling careers, you might prevent the rise of the next William F. Buckley, or other similar pompous rascals. You may even decommission a Bieber! But Buckminster Fuller was also once a boy, and so has every Nimni been. Would you take the risk of never having read, let alone written, *Blueprints for a Sparkling Tomorrow?*

Alas, we would not. Thus the human-flowers must be left to blossom how they must, and be taken care of later. We are now tasked with doing exactly that, but in this case our hatchet is in fact a reasonably-priced academic book. With these *Blueprints*, then, we must critically dismantle the incompossible.

A. Structure and Infrastructure

INSTITUTIONAL STAGNATION

Each metal rusts, just as surely as each man does. And as surely as the metals engages in such rustication, so does each institution. As the dime-store fountain drink cannot bubble eternally, neither can an Executive Board or a Workingmen's Association execute its function perpetually without a Constant Refresh of Purpose. In the same way as ten seconds after one wanders into a room, one is guaranteed to forget why one came, ten years after an organization is founded, it is guaranteed to lose track of why it even bothered to come about in the first place.

The decay of institutions over time is a well-documented truism. Why, just look at what happened to the velocipede, becoming as it did the decadent and functional *bicycle*. Look similarly at the United States Senate, which started out an edification chamber and became a dioxide factory.

How, then, do we keep our institutions rosy and plump? One waters a plant, but one cannot easily water a government. "Yes," the audience replies, "it is very well to say that organizations collapse over time, but without proposals for their uncollapse, what use for ye truths?"

We believe the answer may lie in the elimination of written text.

It is common to craft "mission statements" or "basic rules of orderly governance." This is a mistake. Each set of rules is inevitably a jail cell, each will breed lawyers like vermin. Human life is a shifting multiplex, and efforts to manacle it with policies and promulgations will turn its bulging form necrotic. Writing tames the bulge, thus free-dom necessitates the oral.

But our chief purpose in this segment is not the promotion of oral intercourse. Rather, we are more interested in examining the na-ture of economic and political institutions. Our suggestion that in a reasonable world writing would be entirely banned, is a subsidiary cleanerfish on the main whale of our point. The reader should mainly be learning lessons about how organizations fail to achieve their orig-inal stated purposes.

For who can forget where he or she was on the day the Army was formed? Anyone who insists she was not excited at the time is a contemptible lying viper. Here were a million new gents in uniform, the cream of the nation's teenage crop, ready to spread morality by force abroad and perform elaborate coordinated dance routines at home. Your authors remember sitting in a West Haven bar, still aglow with the high of just having seen the military's first-ever parade, de-bating vigorously (but good-naturedly) with some local longshore-men the question of whether the young soldiers' disposition was best described as "gusto," "moxie," or "panache." Ah, to be a fresh-faced jingoist during those early years of promise!

But just look at our Armed Forces today! Slipshod, poorly dressed, and rife with homosexual debauchery. Now, don't get us wrong, ho-mosexual debauchery is a core component of the life-well-lived, and the highest of the civilized virtues. But is it really why one forms a military? With the obvious exception of the Navy, one may answer confidently that it is not.

The collapse of soldiering into a bloodsport instead of an elabo-rate costume pageant thus shows plainly how institutions can forget themselves. Each needs a regular dousing with the turpentine of re-membrance, or the small blemishes on its exoskeleton will soon find themselves transmuted into implosions.[21]

---

21 Yet take caution from Aphorism #4: *"No revamping of the Institute can hope to achieve the Objective."*

ON THE COMPARTMENTALIZATION
OF SOCIETAL ACTIVITIES

But we have put the horse in front of itself. In order to elaborate a full theory of how institutions act as a restrictive belt upon the expansive paunch of human life, we must first delineate the paunch. We have made clear that life is an ever-flowing thing, like a chocolate waterfall or sewage-hose. Yet in many respects, life is constantly emburdened, squeezed into arbitrary holes like a four-hundred-pound man in a two-inch suit.

Let us now consider the smoking-area, the beer pen, the free-speech zone, and the election booth. Each of these spaces, which we witness daily, represents an attempt to compartmentalize the functions of life: to designate places in which single activities may occur, to the exclusion of less "favoured" ones.

But this futile attempt at atomization ignores the fundamental truth of human experience: life is a flowing blur, and boundaries are both dynamic and illusory. Consider the hand and the wrist. At what precise point does the hand end and the wrist begin? Can a line be drawn with a ballpoint pen? With a chisel? And if we solve that particular quandary (we won't), where does the wrist end and the arm begin?

The point: the hand and wrist are plainly not fixed objects with definable borders like gavels or nations. They are useful conceptions, helpful only *insofar as they apply.* In the same way, the smoking area cannot be said to begin and end, for all objects can be said to either smoke or "be made of smoke." To artificially designate an area in such a way is to inherently call for an appeal to the nature of the space. But can a space have a nature? If one plays baseball on a baseball field is that natural, or deviant? Wouldn't it be just as "natural" to lie on your back and observe the stars, as mosquitoes devour your love-partner? Should we therefore designate all star gazing to occur in and only within the confines of the baseball playground? We think not.

An example of the phenomenon in action: Your authors recently attended a noteworthy academic conference at George Mason University's School of Public Policy. During the process of departing the campus at the conclusion of the event, we chanced to notice a so-called "Free Burrito Night" taking place in the school's Student

Center. Famished as we were after a day-and-a-half of conference-crackers and hotel salads, we eagerly partook. But while the "burritos" in question were on par, quality-wise, with the typical fare of a modern collegiate vendedorium or canteenery, an event supervisor informed us that the products were not to be consumed outside the bounds of the building, and that we must stay within its walls until every last scrap of our meals had traveled from wrapper to tummy.

Naturally, we stood aghast at her transgressive word-edicts, shocked and outraged by this attempt at the forcible spatial limitation of our digestive exercises. This feeble-minded bureaucrat was then courteously informed that the Earth is an organic system, incapable of being arbitrarily spliced into unitary components and autonomous particles. After some minutes of verbal and mildly-physical tussling, we then found ourselves callously ejected from the campus, left to seek our sustenance at a dilapidated and mildly sticky local Shoney's, in spite of having that day given one of the finest lectures on centers and peripheries in Virginia's four-century-long collective memory.

All of this nastiness could easily have been avoided through a light sprinkling of Creative Wisdom. Had the despotic peasant-woman in charge of the burrito table only understood one fundamental truth about the nature of the earthly ecosystem and the social activities of humankind, Virginia's second-best Third Tier university might have had the good fortune to witness a repeat performance of a stellar and prophetic lecture by two well-respected prognosticators at *next* year's Problems in East Asian Economic Development forum.

THE OFFICE AS DISTINCT FROM THE PLAYFIELD
(ELIMINATING THE DIFFERENCE BETWEEN PRODUCTION AND CONSUMPTION)

Related to the problem of whether to play baseball on a baseball field is the need for *eliminating the difference between production and consumption*. The factory-space and the home-space have traditionally been thought of as not only functionally but geographically separate. This is mildly correct, but does not have to be. The incorporation of the factory into the home is not only a possible reality, but an extraneous one.

Let us imagine what might occur if we were to produce all items in the home. Our stovetops would not only create heat, but would create miniature reproductions of themselves that could eventually serve as replacements once obsolescence dawned. Our frigidaires and coolerators might not only chill our eggs and wines, but might veritably *piss them out*. And our greenhouses might not only house plants, but give birth to them. All of this speaks to a wondrous future for the House-husband of Tomorrowland.

How much efficiency is lost in the transition from Home to Workplace to Home again? How about from Home to Shop to Workplace? A great deal, we might suspect, although we simply have not done the research to confirm our prejudice. Still, it cannot be doubted that some benefit would arise from the merging of presently-separate spheres, for life is not the Venn Diagram that its professors often assume it is. If all things were to take place at one place, at one time, we would almost certainly conserve energy, have time to relax for a bit, and save a fortune annually on tires and lightbulbs.

On the Prioritization of the Center Over the Periphery

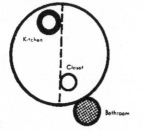

As much issue as we may take with the policies and practices of Marshal Stalin, even we can concede that the Communist movement left one shining idea firmly implanted in the global *Übergeist*. This is the emphasis on the Center over All That Is Not The Center.

The culture of the capitalist (specifically in its modern, Westernized, "attenuated" variety) is a hallway culture as opposed to a living-room culture. It stresses means over ends, principle over purpose. As long as we adhere to certain doctrinal absolutes, the conclusion reached is unquestionable. For an example of the principle in action, consult the topmost portion of the aboveward diagram. As can plainly be seen, it is the hallways that have priority in this mode of living. The actual substance of living, which takes place primarily in the bedrooms and common room, is deemphasized, while the parts of the human house which are intended as mere body-conduits are given great significance. Contrast this with the bottomy segment of the diagram, which illustrates the Communist House, as proposed by Moscow University researchers in 1968.[22] Here the stuff of life is the stuff of housing, and we eliminate the disunity of means and ends that characterizes not only the Capitalist House, but Capitalist Social Relations proper.

This force driving efficient-housing is the same force driving Mr. Lenin's Vanguard Parties, which may have appeared "barbaric" or "grossly undemocratic" to the college undergraduate approaching the man's work for the first time. Driving the communist idea is an emphasis on purpose, and the most direct route to the achievement of said purpose, thereby not only increasing its efficiency-advantage

---

22   Note: the illustration has been translated from the Italian, which was previously translated from the original Russian, thus considerable quantities of information may have been lost in transit.

over its more savage ideological bretheren, but its spiritual fulfillment as well. For while Dr. Capital may leave us wondering why an incorporated enterprise is allowed to reap enormous quantities of profit from the hospitalization and subsequent death of our sainted grandmother, our Red friend leaves us in no such existential crisis, giving us a direct answer to the child's grating cry of "Why?"

Why? So that we may enjoy the common room, without having to travel through miles of needless hallways.

On Failures to Consider the Periphery

Yet for all this Bolshevik claptrap about the Center, we must note that the vast majority of our citizens and lovers reside quite happily in the periphery.

We also note that the prioritization of "destination" over "journey" in practice serves as a handy method of justifying barbarism, and tends to bring about the execution of dissidents and noncomformists, plus further despotic abuses of power in the name of purely theoretical twinkling end-utopias. A life without hallways is a life without substance, sacrificing the joys of the voyage to a mere conception of a common room that will never be reached.

The perfectly contradiction-free habitation-home has neither hallways nor non-hallways, but is an ever-shifting collection of indistinct functions and reporpoised purposes. There is no common-room, there are no hallways, but each thing is all things. In this way, we may both travel and act with ease, while avoiding expensive and complicated identity crises.

The Stoplight and Its Discontents

But in discussing how humans move through their world, and what that world is made of, let us get specific, and probe an example of a specific device that inhibits both motion and essence in destructive ways: the stoplight.

Passive acceptance of the stoplight has persisted for unacceptable durations of time. Its photic tyranny requires immediate jettisoning. We are of the opinion that Humankind can get along just as well without the stoplight as with it, thanks to the coöperative forces embedded within our essence.

Consider the highway on-ramp. In most townships, boroughs, and parishes of the contemporary United States, automobiles join highways via simple one-lane channels, on which no regulation or governance is to be found. Yet the ramp-mobiles and the highway-mobiles marry one another like a zipper, filing in together with unmatched grace and precision. This occurs without any guidance from the forces above. In fact, in those rare crannies of the nation where stoplights have been installed on local Massachusetts on-ramps, we have found the result to be both unnecessary and irksome.

The lesson here is this: Human nature itself can guide automobiles peacefully together at high speeds. If this fact is a true one, and it is, then what purpose can stoplights serve, other than to increase the number of homicides and fender-benders that befall our nation on a weekly basis?

From its discriminatory biases against the colorblind and the impatient to its compartmentalization of human experience into fixed and immovable circles, the stoplight embodies the most corrosive trends in stagnation and bureaucratization. It has proven itself to be a most unholy amigo, and deserves a permanent place on the scrapheap of failed traffic control devices, alongside the notorious and thankfully long since discarded Pigs Walking sign. We therefore suggest the immediate removal and recycling of all known stoplights in the contiguous States, and the adoption of a new de facto mode of travel by which politeness replaces force as the governing intersection ethic.

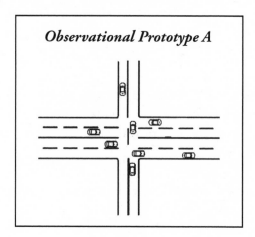

Observational Prototype A

The stoplight only governs us because we invest ourselves in it. If we withdraw our investment, it becomes Just Another Red Light, like the Hot Surface light on our favorite oven or the piercing eyes of an unexpected bat.

Totalitarian Architecture

All architects are fascists. This is elementary, but the interesting question is "How so?" For it is one thing to point out that the architect, like the fascist, believes he can control mankind structurally. It is yet another to account for the fact that fascists wear boots while architects tend not to.

But there are more ways to evaluate a profession than by whether its members wear boots or not. One must look at *practical consequences*, that is to say, that which happens after the other thing. If architects are producing moral horror, we might confirm ourselves by determining whether that which is built *in fact* resembles a horror-house.

How then, does the architect's fascism reveal the naked corpse of his profession? How do we look through its window and watch it undress? How do we trick it into entering its carrier so we can take it to the vet? Let us turn the voice of a practitioner, Peter Eisenman of the Yale Architecture School:

> *They assume that [I'm a leftist], but here's proof that I'm not. You know, I can tell you this: most of my clients are Republicans, most of them are right-leaning. In fact, my client in Spain for the cultural center at Santiago de Campostela is the last Francoist minister. And I have the most rapport with right-leaning political views, because first of all, liberal views have never built anything of any value, because they can't get their act together. I find this public process about what monument we should build in downtown at the WTC site an aberrant one, because since when does the public choose?*

Finally a profession whose most incandescent luminaries see clearly the obstructive nature of democracy and the public. Fascism *gets its act together*, its locomotives consistently rolling up to the platform at the appointed hour and no sooner. We may leave aside the bothersome factoid that the platforms arrived at by these timely trains may have been at the front gates of death camps, as well as the historical reality that no amount of fascism could get a Spanish railway to operate less than two hours behind schedule at all times. It is nevertheless true, on the Eisenmanian account, that popular sovereignty is for dweebs and mediocrities.

But we may meet the charge with a "yes, and what of it?" If one of architecture's most respected contemporary practitioners openly boasts of his seething contempt for the populace and his close partnerships and sympathies with crypto-fascist dictators, what bacon is it off *our* plate? Shouldn't we assess Eisenman not by what is produced by his mouth, but what comes out of his fingertips? The proper thing is not to look at a man's diatribes, for any fool can give a quotation to a newspaper, but to examine his physical consequences. Yes, well enough. Let us witness an account of the structures Eisenman designs:

> *Purposely ignoring the idea of form following function, Eisenman created spaces that were quirky and well-lit, but rather unconventional to live with. He made it difficult for the users so that they would have to grow accustomed to the architecture and constantly be aware of it. For instance, in the bedroom [of the Eisenman-built family home] there is a glass slot in the center of the wall continuing through the floor that divides the room in half, forcing there to be separate beds on either side of the room so that the couple was forced to sleep apart from each other.*

At last we have mounted the apogee of rightist architectural impositionism: the house that literally insists on destroying the marital arrangement through a calculated nefarious design. Further, if we understand P.E.'s political theory correctly, the couple physically separated due to their architect's queer personal brand of whimsy *are not permitted to complain.* Yet complaints to the superior are the heart of this heartless little planet. There is a reason that the national motto "Is there some kind of supervisor I can speak with?" is engraved into every American coin.

There are, consequently, means of making architecture and buildings that do not involve subscribing to Francoism. Place-design can yet be yanked from the jaws of the fascist plesiosaur. Witness, for example, our own proposal for the Motel of Tomorrow:

The Motel of Tomorrow

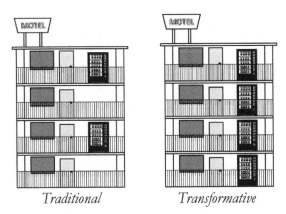

Traditional   *Transformative*

Or look at Venice, or the streets of Louisiana's fabled Crescent City. Places which enmesh their participants in a warm structural hug, that give them the business without also giving them the news. There are more possibilities on this earth than simply death or despotism, at least architecturally speaking.[23]

Rèal Estate

Throughout a human's transient stay on this colourful if underwhelming planet, she will be preoccupied mainly with survival. In clichéd discourse, survival consists of three elements: food, water and shelter. The first two are easy enough to decipher. Without pursuing matters

23 Further indictments of the construction industry's politics can be found in our short book on font design, *Architectural Lettering and the Womb* (MIT Press). In this work, we describe and fulminate against the tendency of the architectural profession to insist on narrow standards for acceptable lettering styles in building designs. We have a theory for the origins of this finicky prescriptivism, and it is decidedly not that architectural lettering has been standardized in order to allow for ease of use and to eliminate potentially deadly error. Instead, we posit the Freudian explanation. Their finely curved B's are the misshapen testes of the architect's primal insecurities, and the houses he builds are manifestations of repressed sexual longing for the Edifice of motherhood. The house is an attempt at structural sublimation, and we need hardly point out what the architect's drafting-pencil represents. (Note that we are under a legal obligation to state that in this footnote, MIT stands for Montana Institute of Telemetry.)

comestible, the torso rapidly withers. Without guzzling the sacred liquids, one is soon destined to parch.

But the third factor, shelter, poses a curious quandary. Where do we live and why do we live there? Instead of answering this question unambiguously, we propose to answer it dialectically, and in the second person. Because it is ultimately *you*, the reader, who will live in the world's houses, your input into the discussion is essential. Through this generously inclusive technique we may discover more that just the outer structure of our dwellings but also the inner structure of your own preference schema.

A scenario: You see an interesting and well-painted house that the local estate agent wishes to sell you. You show up at the appointed time and plod about the grounds, examining its resemblance to the brochure. Upon inspection of the anterior wall, you find that a small segment has been intentionally carved out, and a crystal edifice inserted. You believe in this crystal edifice, forever indestructible; that is, in an edifice at which one can neither *put out one's tongue on the sly* nor *make a fig in the pocket*.

Is it fear that you feel? What is keeping you locked in this item's ominous glow? When it has become impossible to put one's tongue out at it even on the sly, one is surely under the spell of weighty cosmic forces. But what you have found is not that, but this: the Infinite and Unwavering Power of Minor Architectural Variations. The force is neither fairest Eris nor Baron Samedi. It is the logical result of a belief in the potential sway of the Edifice as institution. Reject the realtor, and it ceases to exert itself.

Now, a minor variation on the scenario: instead of a palace there is a chicken coop. It starts to rain. What happens next? Well, I will perhaps clamber inside the chicken coop to elude a wetting, but all the same: *I will not take the coop for a château out of gratitude for its having kept me dry*. Here, the reader may laugh. The reader may even say that in our propounded case it makes no difference - chicken coop or mansion. Yes, say we, *if one were to live only so as not to get wet*.

We have here shown definitively that the Home is both no-thing and all things. After all, Mexico is a country, not a restaurant, and so the nature and usefulness of buildings lives entirely outside of their physical structures, given life only by our interpretations of them.

B. Society, Culture, & Animals

CAPITALISM

C an capitalism survive? Yes, we think it can. For, thanks to it, we now know that every necessity for human-living can be condensed and dispensed in plastic squeeze-tubes, a remarkable achievement. And capitalism itself has proved unexpectedly resilient in the face of numerous attempted floggings. When we, the authors, once had occasion to nationalize a small copper-venting company, we found that no sooner had we distributed ownership of all shares in Righteous Copper Vents & Particles, Incorporated to the masses than the masses cashed out and began to cruise the markets for new and more vigorous investment opportunities. Our brief revolution in collective-ownership was umbilically throttled before it could soak in even its first ray of daylight.

It was not this incident which first poisoned us against Markets and *Marketismo*, but this failure did cause us to think twice about showing up at Capitalism's well-attended annual voodoo soirees (and to proudly rebuff the pointlessly frilly invitation-card we received to the wedding of Capitalism and Communism, though we confess that this stubborn refusal may have been the mere product of 1980's cocaine-fueled mad tempers rather than the deliberate principled snub we would like to have thought it was.) The real turning-point, which has since served us as a very archetype of turning-pointillism, was the moment during the sweltering Seattle summer of 1996 when we realized that true happiness was not to be found in a squeeze-tube.[24]

It is commonly offered that capitalism is the springy mechanical teat from which all innovation suckles. Without capitalism, how would we unleash the Dynamic Human Spirit in all its variety? How would we come up with a useless gizmo? How would we make people pretend to enjoy it? These are indeed questions. But Humanity was not destined to be, and is now attempting to compensate for this error by devouring itself.

Perhaps we might look at things through a differently-bended lens: every time we struggle with a paper jam, or are told off by our workshop superintendent, these crimes should be held accountable. If capitalism can be thanked for our toothpaste, it can also be blamed for the unreasonable price of pistachios. If we bless it each time it hands us forty more ounces of soft-drink for only nineteen cents more, we must curse it whenever our conference-notes are intentionally set aflame by a vindictive airline employee.

The central economic question is, and has always been: are jobs fun? Having used the academy to avoid what is called employment for the large part[25] of our own lives (we attended summer classes through the European Graduate School's *in utero* extension program during our gestation), we are not in a position to comment definitively on whether this so-called "work" is worth participating in. But it has consistently sounded an unattractive prospect.

Consider our own lives: each weekday, we rise between ten and ten-thirty. We lay in bed for half an hour or so, ruminating upon our dreams. Did the presence of a panther symbolize something racial? Something sexual? Can an academic theory be made of it? If so, we note down the theory.[26]

After a brisk latte, and thoughts of going for a run,[27] it is time for

---

24   Elaboration of the insinuated incident can be provided upon private request to the authors. It should not need saying, but the principals of the drama were a potent form of lysergic acid diethylamide and a revelation concerning the way that the impressively large number of commercial toothpaste brands available under capitalism inexplicably failed to bring about ecstatic spiritual wholeness.

25   all

26   See: Nathan J. Robinson & Oren Nimni, "The Panther as Racial Ontology: Feline Signifiers and the Academic Imagination," *Journal of Oneiric Scholarship* (2014).

27   If we decide not to go for a run, we nevertheless "identify as someone who has gone for a run." Identities do not depend on facts about the world, but on individual subjective

some morning reading. First the classics, then the news. Some time with Marcus Aurelius's *Meditations* may yield further useful insights for potential scholarship.[28] Next, we must form an accurate, unbiased picture of the world around us. So, naturally, we take *The New York Times*.

At one p.m., we co-teach an undergraduate seminar on Utopian Torts. This does not require any preparation.

Next, office hours. To maximize efficiency, each student is given the same advice. This is possible, because all human beings are the same shape. We are all more alike than different, thus there is no need for a professor to provide "individualized" (i.e. divisive) feedback. Thus, we tell all students to improve their attendance and use longer words.

By this point, it is five. Having put in a full half-day's labor, we are due to go for an afternoon pastry. If we are feeling especially committed, we will write an article in the time it takes to eat the pastry.

But where were we? Capitalism. Yes, capitalism cannot and should not be sustained. For the capitalist is at once an atheist and a be-theist, a negation of simultaneous living and nonliving religious selves. The capitalist believes the quadrillion miniature interchanges of a functioning market will produce the good, yet simultaneously refrains from voicing a conception of the good save that which is produced from said aforementioned interchanges. What kind of catastrophic moral nonphilosophy do we even have here?

The capitalist and the nihilist are not the same man, but they do sit side-by-side and sweltering in the same metaphorical sauna. *The capitalist sees the nihilist nude*, so to speak. Or is his own nude self in reverse. Or has a refracted non-nudity that can only be negated by the very fact of his initial nudeness. Or defines an abstract anti-nudeness that distinguishes itself wholly from the nakedness of the nude within. Each of these formulations accurately describes the same phenomenon, with escalating levels of precision.

But where is the boss? Everywhere, so it seems. For it is often forgotten that capitalism is not just the cart that carries our goods to

---

perceptions. By thus identifying, we cause ourselves to have exercised, even when we have not. We can testify strongly to the results yielded by this strategy, which has caused us to identify as extremely fit.

28   See: Nimni & Robinson, "Pretending to Have Read Marcus Aurelius as a Strategy for Luring a Mate," *Bachelor's Monthly* (2013).

market, but it is the man atop the cart whipping the horse to hasten it. And though we have never counted ourselves among those who harbor irrational sympathies for equine suffering,[29] we must concede that humaneness and horsewhips contain contradicting values.

Is it conceivable then, to have capitalism without having constant whippings of underlings by superiors? If such a paradise is possible, we have never seen it advertised at a conference. Though it might be *desirable* to live in a whippingless world, the desirable and the feasible do not always make love on the same deckchair.

Capitalism, then, has a central flaw. No matter how many barrels of toothpaste it successfully drowns us in, for every gallon used to *brosser les dents*, another will mistakenly drown a kitten. That is to say: the profit motive makes death inevitable. For life is not always profitable. Occasionally it is a downright catastrophic financial proposition. Yet if we depend on life *selling well*, we may find ourselves tossed out like yesterday's outmoded handbag. *There is no word for "love" in Morse Code*, and there is no sense of the sublime in a market economy.

Thus: the first and most essential prescription we therefore offer is this:

Stick a worm down the collar of Capitalism.

The wording is calculated. Having seen what revolutions give us (Robespierres, Trabants), we know that we shall not displace Capitalism toward any especially spectacular end. You've got to break a few eggs if you want to have quiche tonight, they told us, but we only ended up with yolk in our hair and albumen down our trouser-leg (quicheless, too, naturally).

The key then, is not to displace capitalism *per se*, but to create the conditions under which capitalism may be *ameliorated* and shall ultimately wither. We shall slowly unbutton capitalism, draping a suggestive fingertip along her bosom, never letting slip that we only wish her to disrobe so that we may set fire to her undergarments.

---

29   See "Horse-Centrism and the Decline of Values," in Part II, p. 61.

GOD

But capitalism is just one of the Throbbing Monoliths that direct contemporary life. Let us consider another: God. It is by now plain to all that "God" as an entity has ceased to exist, for if God existed he surely would not tolerate the number of intellectual raspberries blown at him on a semi-regular basis. What kind of sad ditherer would sit on his toadstool and let little blasphemists prod him in the toes, when righteousness and good manners call plainly for vigorous smitings? The decline of the smite logically necessitates a non-god, or a God slipped so far into desuetude as to be indistinguishable from his impotent natural counterparts (sun, moon, waveform, etc.).

The interesting question for us, then, is not *whether* there is a God. Even such a God as *did* exist would have to be so cruel as to be nonexistent. Rather, the most engaging of the Goddy and Goddish questions is as follows:

Is God too big to fail?

Let us address the question through parable rather than fact. Our cousin Peter J. Robinson-Nimni[30] was born with an enlarged heart, yet this did not prevent his death from heart failure at the age of 23. It seems plain that there is no correlation, whether verse or inverse, between largeness and failure.

Yet lest we be accused of attempting to substitute the aberrant heartitude of an obscure and unloved academic/familial relative for The Cold Hard Unforgiving Sand-Blasted Concrete of Empirical Fact, here are at least four other anecdotes about big things that have not failed:

♦ The ego of one Slavoj Žižek appears sadly to suffer from no upper size-limit, yet his work remains inexplicably critically-acclaimed.

♦ Ted Turner International Airport in Atlanta, Georgia is the largest such *puerto* we have ever found ourselves docked in, yet it seems to run with relative smoothness.

---

30   a.k.a. Nimni-Robinson

- The sexual prowess of Mssrs. Robinson and Nimni is as vast as vast can be, yet they have never yet failed in their bedroom endeavors.

- The 1976 Oldsmobile Custom Cruiser is reported to have been the largest station wagon ever built, weighing a supposedly unwieldy 5 metric tonnes, yet ours has never let us down.

- The George Bush Airport in Houston, Tejas is even more enormous than its Turnerian counterpart, and yet George Bush still manages to be an unprecedentedly enormous moral monstrosity.

It is clear then, that God is here to stay, however unwelcome he may be in our beds. But what do we make of the looming presence of this ungentle giant? Ought we to worship him? To leave a buttercup on His windshield? To sacrifice a Congressman in His honor every fortnight? It is as unclear how one is to react to God as it is which God to disbelieve in the first place.

The answer may lie in theology. Once during graduate school, when one of us was briefly in love with a theologian, we picked up a bit of religious knowledge in order that we might impress her, and though we were cruelly spurned on the very night of our dissertation defense (a dissertation ironically devoted to the question of whether religious questions influenced earthly romances), we retained from the incident a dollop of Christian trivia that we continue to haul out at vicar's teas and Nimni family baptisms. Allow us, then, to quote Augustine:

> *"First of all, we would ask why their gods took no steps to improve the morals of their worshippers. That the true God should neglect those who did not seek His help, that was but justice; but why did those gods, from whose worship ungrateful men are now complaining that they are prohibited, issue no laws which might have guided their devotees to a virtuous life? What could have been on their tiny minds, or if so, who would dare say that he truly knew the outcome by heart and could swear to it in an emergency? Surely it was but just, that such care as...*

...men showed to the worship of the gods, the gods on their part
should have to the conduct of men. But, it is replied, it is by his
own will a man goes astray. **Who denies it?"**
                                        - St. Augustine, <u>City of God.</u>

Indeed, not us. But the *theological* (as distinct from *logical*) question is
whether that sinister little pear-thief should be given any credence at
all, if pears are given their proper moral weight in the assessment of
speakers' motives. It is motives, then, and not God, with which we
must chiefly bother ourselves.

But let us digress on pears. It is because we have consistently identified with the pear that we so distrust Augustine.[31] We like the Pear
because it makes those of us with pear-shaped bodies feel good in
our own skins. "Ours is a natural form!" we can cry to the fun-pokers.
Plus, any man who has ever fed a horse a pear out of his hand, and
had the horse dribble little gooey pear-guts all over him, knows that a
pear can be, in its finer moments, resplendent.

Where does this leave us on God? It is this: God exists, *but is*
incoherent. To affirm or deny the existence of an object requires an
understanding of what manner of object we are affirming or denying.
God is not any such sort of object, thus God can neither be affirmed
nor denied. He can only be taunted.

Slavoj Žižek: Academic Charlatan

Which brings us to Slavoj Žižek, the present-day darling of the bearded academic regency. Žižek may well be incompossibility personified.
Certainly he is a more insidious creation than *The Guardian* implied
when it labeled him a mere "serial peddler of false histories and a
manic creator of pamphlets."

Alas, first: a note regarding our own biases on the matter: Our
intellectual rivalry with Žižek and his minions has a bit of a history.
From an early 1990's exchange of savage op-eds in the pages of the
American Journal of Trades and Tradesmen to Žižek's rude and unprecedented walkout during our presentation at Seattle University's 4th An-

---

31 See L.G. Ferrari, "The Pear-Theft in Augustine's Confessions."

nual Conference on Academic Reboldening, the Žižek-Nimni-Robinson relationship has been a triangle of base hatred.[32]

We do not know which particular spectacular opinion of ours initially set this war alight, but we do know that over time we have somewhat ruffled Mr. Žižek's already-unruly beard. And rightly so, too, for in his consistent hawking of nonsensical philosophies to the young he resembles nothing so much as the aging schoolyard pedophile, handing out penny-candies and erotic postcards to the children at in exchange for depraved favors behind the bicycle-shed.

But to turn to the muffin-shaped man himself. Has one coherency been issued from beneath beard across its travels from Ljubljana to Zuccotti? If lips there indeed are beneath those hairs, from them have we learned a *trifling damn thing* about who we are and what in the Devil's name we're supposed to do about all this? To ask the question is to scoff at it.

In his own words, Žižek "cast[s] violent doubt upon [anyone] who does not stand in admiration of the storied professor who, offered by his genie one use of the time machine, chose to return to France and re-behead Louis XVI. *An unnecessary necessity.*"[33]

What iz Žižek thinking here? It is an elementary maxim that *all* necessities are unnecessary. If he expects his reader to slip beak-like over the logical fallacy, he has not counted on the "Nimni eye."

---

32   Save for the Nimni-Robinson leg of the triangle, which has been relatively amicable. For further explication of the relationship's nuances, please see the below visuo-triangular diagram:

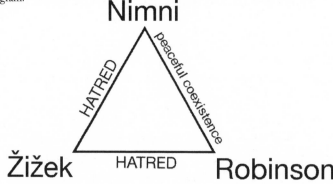

If the point remains unclear, we recommend the reader consult our now-notorious guest column from the *Vancouver Sun* entitled "Isosceles Loathing: The Geometry of Nemeses."

33   Slavoj Žižek, *Bleeding Eye: Lacan Through the Violent Lens,* Verso (2001), p. 784.

Žižek continues needlessly:

> *When people say 'navel-gazing,' this is untrue. Film is a navel* **through which** *we gaze. What matters is what is on the other side. Just as the porthole does not make the ship, the fact that Hitchcock made the film does not make it Hitchcockian. To shift from the man to the tendency in this way requires a reversal of identity;* **that** *is ideology. Time has a secret hidden in her vagina.*

What kind of monster says this? It is true, of course, that Hitchcock possessed a navel. But watch Žižek's sleight-of-hand as he conducts a shift of his own, turning navel into porthole and porthole into unsubstantiated conclusion. If we were to peel away the thick rhetorical coating Žižek has slathered on recklessly, we would see that ideology does *not* work as it does on a ship, where there is first and foremost a captain.[34]

His crimes only reproduce when we turn to the act of political prescriptivising:

> *Did Robespierre go far? No, I claim Robespierre did not go far enough. Terror, in its purest form, decenters. Did Robespierre decenter France? No? Then he did not terrorize it. I am reminded of the old joke where the peasant woman is being brutally interrogated. "Are you afraid?" says the official. "No," replies the woman, "I am terrified."*

This is simply poop. *We all know* Robespierre did not sufficiently terrorize France. That has never been the point of the exercise. What matters is under what *aegis* he did so. Žižek somehow believes it is the center that decenters rather than the opposite. How can a philosopher, of all people, be so out-of-touch? (Later, through an act of mishearing, Žižek has turned our theory of the Big Udder into something else entirely.)

Any one of these crimes ought to justify the tearing up of Žižek's Scholars' Union membership card. Yet he has continued to belch

---

34 We would be remiss, however, if we did not here note Aphorism #5: *"No man looks through the porthole without in some sense getting wet."*

poisonous philosophical fumes onto the magazine-stand and best-seller-lists with an infuriating regularity. That the police permit this, and have not yet gone rooting around in the hemlock-drawer, is a

DIAGRAM 1

Slavoj ← Žižek

testament to the depth of the post-Grecian decline in civilization's capacity to deal with the bearded and obscure.

There are many equally sweaty with whom we would happily share a globe so big and blue. But the admission price to our parlour is clear thinking and rigorous courteousness to one's enemies, and so the fat hack Mr. Žižek must forever be banished to gaze through our glazing and pine upon our stoop.

And so, if the question is "How are we to Žižek ourselves?" The answer is that *we are not*. We are not in the least.

AWAY WITH THE INTERESTING

That our bestseller lists remain dominated by Žižeks and Gladwells is to be expected, however. Populist hacks will triumph until we rid ourselves of the notion that *readability* is an acceptable criterion for judging scholarship. Their success speaks to a mass delusion that is being passed around: the delusion that a thing takes on value to the extent that it is *interesting*, and loses value to the extent that it is not. But some important things are very boring. For example, who would want to learn how to design and operate bridge removal machinery? Yet how many of us would be happy if bridges were left to proliferate wildly across our rivers, reproducing by the thousands and sitting disused and uncrossed? If we chose only ever to do exciting things, we would spend most of our time waging wars and hardly any of it attending conferences. Yet which is more crucial to the reshaping of the geopolitical order? Anyone who has attended an annual meeting of the American Sociological Association would surely hesitate little before exclaiming an answer.

The problem with the interesting is that it is biased towards that which interests us, and we are interested in some very unwise things indeed. Watching an orphanage explode is interesting. Filing the tax exemption paperwork to classify one's office as an "orphanage" under an obscure state statute is comparatively less so. The things we must do are not necessarily the things we would watch ourselves do on television, even though a television channel dedicated to broadcasting nonconsensually acquired images of ordinary citizens undergoing their daily defecation rituals would likely achieve an impressive audience share.

Our own work has set out deliberately to correct the bias toward the compelling. If the reader is interested in us, we have failed our task as scholars. Nobody should *wish* to read the American Journal of Sociology, for to be popular is to pander; it is a small step from the introduction of readable content to the appointment of Michael Bay as guest-editor and the ending of each article with an exclamation point (!).

The more unread a scholarly article, then, the more integrity its writer maintains; for each additional eyeball on a page, one must pose the wary query: "Yes, but *what low textual deeds did you do to obtain said eye?*" Nobody builds an eyeball collection without burgling a few caskets, after all. For the academic interested in staying true to himself, then, book sales are like golf scores, success being measured by the depth to which the skilled driver can make them sink rather than the heights to which the flailing putter can make them rise. By this measure, we ourselves are the twin Jack Nicklauses of contemporary academic publishing.

The only way to free ourselves of an outbreak of Žižeks, then, is first to free ourselves of the dangerous notion that to be interesting is a value of itself. Goebbels was interesting, so is a mudslide. But purity of thought requires speaking only to oneself; to socialize is to compromise, to sell books is to sell oneself, and to sell oneself is to whore the intellect, with each book copy sold being a new act of analingus performed upon the consuming public. Thus we cannot resent our colleagues who successfully sell their manuscripts, for we know their generous monetary advances and celebrity-ridden release parties serve only as proof of their shameful professional debasement.

THE PSYCHOLOGY OF MODERN LEFTISM

These recent ominous upsurges in book sales and popularity are symptomatic of deeper dysfunctions in leftist psychology. Of course, Che was a sellout, licensing himself for all those T-shirts, and we ourselves would never have done this. But the rot rots further. Consider this printed advertisement, ripped forcibly from a recent edition of the *Taxpayer Service Card*:

[advertisement redacted due to licensing restrictions]

Is it nonsense? Or supersense? It certainly caused *us* to reconsider the renewal of our subscription to the *T.S.C.* No number of intriguing featurettes on the latest developments in 501(c)(3) regulations can compensate for an advertising department that considers the use of chatty colloquialisms in printed copy appropriate or even "charming." *A professional news organization with a professional readership should know better than to intellectually debase itself in the greedy hope of scoring an additional impulse-buy or two from subway delinquents beholden to the latest modernist fads in advertisement-text.*[35] For the love of Fuller, **tell us what you are selling and why we should buy it.**

This crotchety clarion-call for clarity behind us, we can proceed to the point: The rise of the Left has run parallel to a corresponding rise in Obfuscation. Young Mr. Postmodernism and his brother-in-law Dr. Deconstruction have sapped us of our will to mean, and informed us that everything is acceptable. *Everything is not acceptable.* Some things, such as the deliberate impression of bootprints upon an infant child, are wholly unacceptable. They should be reacted to with withering glares.

As has oft been wisely noted, "descriptive linguistics is the gateway drug to democratic nihilism,"[36] and while we generally dismiss the notion of gateways as a heap of Orientalist consequentialist horseshit, in this particular case the statement is somewhat sound. After

---

35 Aphorism #6.

36 A point we have gently lifted from something Prof. T.W. McCarty once said about the connection between tax policy and WAR (the one of "Why Can't We Be Friends?" fame rather than "Dear Prussia: we cannot be friends, love Austria-Hungary" fame.) This footnote should in no way be seen to constitute an admission of plagiarism.

all, it must be remembered that even the deepest valleys of horseshit once sprang from a physical horse.

For it is true, is it not, that if one does not believe grammar can be prescribed, one must believe all values to be arbitrary?

We cannot see how one could conclude otherwise without holding at least two self-contradictory opinions. If values spring from heaven, instead of from the anus of each mortal consumer, then both grammatical and moral prescriptions hold. If values indeed anal be, it is more difficult to see why schoolchildren should be punished corporally each time they muddle a tricky subjunctive.

Thus: if we are to avoid moral chaos, we cannot blur our grammars; if a characteristic is mutable, it *will indeed be muted*. In *Blueprints für einen Sekt Morgen: Gedanken über die Einrichtung einer amerikanischen Reich*, the present work's Teutonic translatomus, we seem to remember establishing this concept as the proverb of our time, which therefore necessitated scrupulous and many-sided (polyamorous) investigation. We must therefore give it a minimal nod here, and dismantle the aforementioned nod into its component parts, in order to expose nihilism as the nude dancing postmodern lady it has lately become.

We have chosen to react to the abhorrence of contemporary academic nihilism with an equally vigorous Fiery And Unshakable Moralism. Thus far, it is proving extremely pleasurable; you ought to try it someday when you have a blotch of free time in your pocket. Simply lay out a set of Values (For instance: "Marxism is a crate of barbarous codswallop" or "The suffering of the weak shall not be tolerated") and shower with rage those among your contemporaries who fail to live up to them. Though it is not guaranteed to acquire you friends at art-parties or gumbo nights, it is promised to give a release and even perhaps a refreshing sigh.

The Left is in crisis. This is agreed even by the Left, it is belted loudly by all of their birthday-party Chomsky impersonators and streetcorner frown-merchants. How has the Left shifted from waving flags and megaphones in exalted victory to littering Brooklyn with the same poorly-typeset pamphlet over and over forever until eternity? *What comes after farce?*

The Left has lost sight of its assets: the hector and the tut. It is time for us to get serious about superciliousness, and let sentimentalistic relativism fall from the party yacht. We must cease to respect those values other than our own, and must begin to craft a vision of the just society that extends beyond having a fervent desire that people will come up with visions of the just society. Only this way can the left brain be made right.

ORWELL'S NIGHTMARE

Let us burrow somewhat further into the spirit of the ideal leftist. It is commonly supposed that the *Cheers* bar is Orwell's Nightmare, because *everyone knows your name.* But this is not so. Nor is it true that Orwell's Nightmare is for the word "Orwellian" to have come to mean precisely the opposite of what Orwell most stood for. Instead, turn to the man's biography:

> *[Orwell's] proletarian affectations in the BBC staff canteen-- slurping tea from a saucer and rolling shaggy cigarettes-- embarrassed colleagues and shocked the doormen. Friends were struck by his peculiar combination of gaiety and grimness...*[37]

To fail to slurp one's tea: this, then, is what Orwell feared the most. But let us dwell on this notion of the "proletarian affectation." We reject it. For our proletarianism has never had a lick of affectation. When we greet our secretaries as "comrades," we mean it honestly, with no sense of irony. When we tell our mechanic that he is doing valuable and noble work, that to work with one's hands is the wellspring of personal dignity, that he is doing work we only *wish* we ourselves could do, we are being perfectly sincere. When we ask our barista why he has not unionized his colleagues, we genuinely do not understand what he is doing with his eyebrow in response.

For what is the rôle of the intellectual? The rôle of the intellectual is to offer an inspirational example to the working classes, that they might observe, learn, and rise up. If we spend most of our time in idleness and pastry-consumption, it is not because we are insufficiently committed to the transformation of human society, but if anything because we are *too much* committed to it. We are *being the change we wish to see in the world*; we wish for a world in which all human beings live lives of leisure and dilettantism, thus we engage in leisure and dilettantism.

---

[37] Jeffrey Meyers, *Orwell: Life and Art*, 209.

To live the academic life, then, is the only route to social change. If one is unwilling to *fully* inhabit the university, then one's commitment to revolution can only be partial and tentative. But because we respect the workmen who serve us (whose names we make sure not to learn, because of the aforementioned first part of Orwell's Nightmare), we serve their cause through our lifestyles. We leave no gratuities, because tipping makes capitalism appear tolerable and humane, and thereby enables its perpetuation. And we are consistently rude to cleaning staff, in order to expose the oppressive class dynamics inherent in the relationship. (To be kind to cleaning staff is cruel, for it gaslights the poor wretches into feeling as if this terrible unjust system must somehow be gentle and well-meaning. Those who are nice to waiters do them a far greater disservice than those who denounce them to their managers for failing to provide adequate quantities of saltines *avec le chowder.*)

Thus: reject Orwell's nightmare in your daily life, by adopting the Gandhian philosophy of change through self-improvement.

HORSE-CENTRISM AND THE DECLINE OF VALUES

But since we are discussing ideal values, we should also explore their opposite: the horse. As a prerequisite for further perusal of this chapter, please imbibe the following dispatch issued by the British Broadcasting Corporation:

> *The braiding of horses' manes is being used as a code to mark the animals for theft, police have said. Following a warning from Fife Constabulary about 10 cases have been reported in the past couple of weeks. Pc Ian Laing said: "It probably happens throughout all of Scotland. It is certainly a problem throughout England." He urged horse owners to be on their guard and to report any braiding found on manes or other suspicious activity. Officers said a variety of horses had been targeted in this way and it was not only expensive thoroughbreds which were coveted by the horse thieves. P.C. Laing also recommended that all horse owners register on the HorseWatch Scotland website for more local information about, and better local protection from, equine crime.*

The recent spate of equine crime has alarmed us, and must surely alarm you also, gentle reader. However, it is a demonstrable fact that horse-thieving does not simply spring from the ground like a spud or clover. Nor can equine crime be blamed solely upon equine criminals. Rather, each is dredged from a common poisoned culvert, namely the ubiquity of horse-centrism in modern American value systems.

For too long the horse has stood in the popular imagination as a symbol of triumph rather than one of shame. Horses have now been painted on nearly every gas-station wall in the contiguous states, and odes to them increasingly deluge undergraduate poetry workshops. It is now a rare politician indeed who has never ended a speech with that timeless cliché: *the arc of the equine universe is long, but it bends towards horses.*[38]

Yes, yes, they're cute. Any man who has ever fed a horse a pear out of his hand, and had the horse dribble little gooey pear-guts all over him knows that a horse can be, in its finer moments, resplendent. Yet this temptation to adore the horse is precisely the problem. Horses and hedonism go hand-in-hand. He who admires horses loathes himself, and must satiate himself with pleasures of the senses in order to stave off the mounting reserve of guilt that is always in danger of bursting forth.

We therefore propose a substitute outlet for humankind's affections: the arthropod. Anyone who has attended a lobster wedding knows full well the kind of profundity and romanticism of which these divine creatures are capable. Yet the arthropod languishes in America's batting-cages and seafood joints, stripped of its potential and dismissed in its attempts to make edifying contributions to civic life.

All injustices must eventually cease. But cessation comes only through action. Burn the horse and anoint the arthropod! It is up to you, gentle citizen, to personally find and replace every Horse-Themed monthly calendar with one dedicated to Festively-Costumed Lobsters and Crawdads. Every moment that you are not actively opposing horse-centrism, you are tacitly consenting to it.

[38] Aphorism #7.

CONDEMNING THE ARTS TO THE FIRE[39]

Horses and nihilism are emergent consequences of a barbaristic cultural order. But culture has its other poisons, foremost among them art. Our position on art has wavered from semester to semester, but this year we find ourselves strongly opposing it. The central problem appears, to us, to be that the arts that are beautiful are not useful, and those that are useful are not beautiful.

After all, what have the visual Arts done for you lately? Zilch! They are unlikely to have healed your marriage or dismissed the twenty outstanding formal complaints against you filed by your students. They may have sizzled your cones and dazzled your rods, *but that is about it.*

Thus, the best thing to do with the Arts is to toss them into the flames and warm yourselves by the heat of their combustion! Can a painting sing you a song? No! Can it pass crucial civil rights legislation? No! Can it keep you cozy on a cold winter's evening? Only if it is aflame! We have never met an artwork we would not consign to destruction.[40]

Art is not rust, but at least rust transforms over time. It waxes, it ebbs, it shifts. It is dynamic where art is static. During artistic creation, the artist creates motion (and therefore life) *only so as to not have to create motion in the future.* Once the "work" is completed, it is immobile. This makes art anathema to Healthy Living. It is framed and therefore compartmentalized, it is structured and therefore restricted, it is motionless and therefore dead.

In *Discourse on the Scale of Art*, Hume posits a simple, though complex, quandary. Why is it that so many paintings are found in frames? Would we not expect, if art were made rationally, that frames would have little to do with it? A corollary question invites itself: why is art generally of a uniform size? Yes, one has everything from the painted

---

39 Textual note: In an early edition, this chapter was misprinted as "Condemning the Ants to the Fire," an erratum that painted us as two sorts of magnifier-wielding school-age miscreants. We repeat here, for those who have only recently traded in their creaky 1st editions for gloss-drenched new ones, that in fact we favor ants. Favor them very much indeed. Ants are nature's jellyfish, and anyone who has a significant problem with them has a minor problem with us.

40 This includes the painting *The Tilled Field* by Joan Miró i Ferrà and *Periscope* by Jasper Johns.

thimble to the four-story balloon dog, but on a *cosmic* scale the distinction is negligible.[41]

Art inhabits the museum, we go there when we wish to be seen gazing at some of it. But why does art choose to live *there*? Could it not equally well live in the pharmacy or the privy? It is worth undermining the formalized nature of the "collection," which says that the art is *here* rather than *there*, that it has been gathered from the wild and placed on display.

For when we build museums, we implicitly concede that the world is not beautiful. "Here are the things worth looking at," we say, setting up an opposition between the attractive things within and the abominable things without. Under this approach, our exhibit of classical vases will be sublime, but our freeway overpasses will be a blight. But how can the difference be justified in principle? Should not every overpass be a vase? Why do you see so few frescoes on freeways? It cannot purely be blamed on the transit bureaucracy, there must also some convenient social delusion about the origin and function of art.

Art has always been an odd birdfellow. Nobody knows what it is, yet everyone claims to have formed an opinion about what it ought to be. Nobody can tell you where the art ends and the sewage begins, yet nobody trapped in a standpipe will claim to have witnessed a masterpiece. Hence the emergence of that schoolyard paradox: "There is art in the Louvre, but is the Louvre art?"

However, it is not necessary to decide which mummified animal carcasses are art, and which are worthless kitsch, in order to observe the central arbitrariness of the whole art notion. "Art" is short for "artist," but that's not the point. The point is that no object should be made that does not have artistic worth, and the idea of keeping a painting in a frame just so that it doesn't run away misunderstands the difference between animate and inanimate objects.

It is our opinion that any art worth having must sway, slide, and jiggle, must be the size of atoms and of canyons, must break free from all enclosures or frames, and must simultaneously reflect and create the means and modes of human living-patterns. *Our current Art, meeting none of these criteria, must be destroyed.* Dispose of the printed word, excise the painted word, and find peace.

41   See O. Nimni/N. Robinson, "Why Is Art So Small (and So Large?)" *Rose Art Museum Seasonal Catalog* (Spring 2009).

C. Political Arrangements

EXPANDING THE REALM OF THE POLITICAL
BEYOND THE PRESIDENT

The United States has historically had a fascination with electing "presidents," tall men who strut about issuing pronouncements. These man, innocuous though they may seem on a postage-stamp or hanging in a gallery, are in fact extremely powerful, and do incalculable damage to the cause of Universal Joy.

The core problem is that President *seems to think he's the President.* We have noticed this with increasing frequency over the last several days. Yet this man is *not* the President of vastly more things than he *is* the President of. (He is not for example, the President of the Emerson Electric Company or the President of our hearts.)

Let us consider the effects of the inauguration of this new President of ours. Let us pose a question: Aside from changes in discourse, did you notice a solitary change in the essence of your day-to-day existence during the transitioning from President to President?

Here is the thrust of our point: Custom has greater power than law. *Law is but a series of magic words.*[42] We are guided far more by tradition, expectation, and habit than by the man in the enormous chair with the unusually-shaped office. The arms by which the members of a society are to be kept within their duty can therefore be exhortations, admonitions, and advices. We have no need of this President or his supposed "policies." Apart from those whom he has imprisoned or betrayed, none of us has any reason to pay attention to this man's

---

42 Aphorism #8.

mad ravings. Let us treat him as if he were a mentally ill uncle, who rules an imaginary empire from his balcony. Perhaps some local lads from down the block bring him food and water, and enjoy pretending to be his Armed Forces. But at the end of the day, no amount of power he manages to achieve, no number of stray cats he manages to marshal for the protection of his realm, will legitimize him or give his bizarre edicts and scrawled manifestos any moral authority.

The realm of the political, if the political is defined as deliberation over the uses of power, stretches so far beyond the realm of government as to make government seem almost inconsequential. Day-to-day living is governed far more by the politics of the family, the workplace, and the relationship than by the actions of legislators or presidents. Let us re-channel our discussions of "national politics," and instead discuss the intensely local and personal matters which govern us more directly. We must spend time figuring out the politics of our jobs and of our friends, rather than allowing this lunatic uncle of ours to define the subject matter for our consideration. The United States Congress only matters because we believe it does! Stop believing, and it disappears!

JUDGES AND
THE MASS PRODUCTION OF MINIATURE GAVELS

But presidents are not the only law. Judges, too, show up occasionally to overturn a civil rights act or demand that some ne'er-do-well be lethally injected. Judges have even been known to hold well-regarded local professors in "contempt of court" for the mere act of attempting to show how their so-called "parking tickets" are made impossible by the fallopian nature of parking-structures. What does Law cook up in her cauldron? To judge by the state of the judges, *nothing worth imbibing.*

American jurisprudence has historically been clouded by the biases engendered by its arbitrary symbols of authority. The robe, the gavel, the enormous oak bench from which the judge can shoot piercing stares into any poor soul at his leisure. We have thankfully disposed of the Wig, but its progeny remains intact in the form of the so-called "judicial toupée" sported inconspicuously or conspicuously by the more infirm and paunchy of our legal professionals.

These symbols of authority, each with its own subtle psychological effect, have skewed the American justice system to the point where it is impossible to recognize whether a particular verdict came about as a result of a neutral application of the law to the facts or simply a jury's subtle persuasion by the shimmer of a newly polished gavel.

We therefore propose that conventional "foot-long" oak gavels be abolished, and replaced with small, novelty plastic gavels. These may optionally make a "boink" sound when used. These new gavels are to measure approximately the width of the judge's left index finger, so as to most thoroughly undermine the legitimacy of the court.

Similarly, we remain in strong support of the replacement of traditional *noire* robes with tie-dyed ones. In addition, while wigs may remain white upon request, when the wind reaches these crowns at appropriate velocities, wig movement should meet or exceed 5 inches. The flowing white locks of justice billow majestically in the breeze.

CAKES AND THE OBSOLESCENCE OF LAW

Aphorism: *At a certain level of cakes, law becomes obsolete.*[43]

The irony, of course, is that this maxim is *itself* a law, regardless of the number of cakes present in a given system. But despite the internal tension, the truism holds. Law and cakes are not complementary, as was once thought. They fall along a well-defined continuum, for those who are given no cakes must be disciplined through law, and those who are well-stuffed with cake have no need for legal controls.

This is the very reason Marie Antoinette necessarily had her *tête* forcibly segregated from her *corps*. She understood well the power of cakes, but flatly ignored the continuum. She desired a world with both cakes and law, a contradiction that could have been avoided had she paid heed to Lincoln's timeless admonition: a horse divided cannot stand. She is, of course, to be praised for not wishing to *have her cake and eat it too*, for the violation of two ancient edicts by one 18th century monarch would be too much for even a Universe as indulgent of Mother Absurdity as ours to bear. But wishing to have your cake *and a system of legally-enforceable restrictions on behavior, too* is almost as heinous a crime in the eyes of Logic.

---

43 Aphorism #9.

THE LEGAL MIND

We are opposed to law in all forms, and the lawyers we have met (and who co-teach our undergraduate seminar on Utopian Torts) have consistently been tedious company at parties. Among our fundamental principles is the belief that Human Be-ings shall serve no gods, no masters, and no arbitrary system of rules. There is a reason that "lawyer" and "liar" have the same pronunciation. AND YET! Something about the mind of the legal professional fascinates and arouses us. Thus, while generally legal practice is among the basest of evils (just as strontium hydroxide is the evilest of bases), we find something minute yet un-nonexistent to praise about the way that the Lawyer thinks, if not acts.

What does the lawyer do, other than connive? The lawyer, in her purest form, is a Reasoning Machine, a black box into which premises are placed and from which deducted truths are spewed. Astute readers may respond that such a person-contraption seems at odds with our philosophy, which tends toward anti-reason. But there is a certain romance to the scientific approach to truth. The task of the pragmatic utopian, then, is to save what is worthwhile about being reasonable, while never allowing ourselves to thereby begin mistaking statutes for justice.

But what is it we are talking about when we talk about laws? Dr. Chomsky hears our inquiry and replies as follows:

> We are asking--if we are serious--whether the law is a sufficiently precise and delicate instrument so that it can label a monstrous crime as a violation of law. Similarly, in considering the legality of the intervention itself (apart from the means employed), a person who is serious about the matter is not examining the propriety of the act but rather the adequacy of the law. Suppose we were to determine that international law does not condemn the United States intervention as criminal in the technical sense. Then a rational person will regard the law, so understood, with all the respect accorded to the divine right of kings. For Reasons of State, p. 19-20.

Let us *not*, then, be nuisanced into assent. The *mussolingua* that comprised the language of Italian fascism would erupt in mirth at a witnessing of today's legal tribunals, in which it is *people* rather than law who are on trial. Much as a menacing and shout-laden phone call to the editorship of the *New Left Review* was in order when the letters in "Oren Nimni" were mistakenly and misprintingly arranged to spell "Ovren Minri" in a byline (despite the universal consensus of colleagues that such use of the telephone was misplaced and should be avoided), the inversion of this notion of *what precisely is at stake* should be met with screaming worldwide hissy-fits until the error is corrected. That is to say: stubbornly oppose all laws, until laws there are no more.

TACITLY CONSENTING TO TACIT CONSENT

Of all the things we have managed to tacitly consent to,
tacit consent may be the most insidious.
-Nimni/Robinson,
Forgotten Classics: An Anthology of Unpublished Texts

Law has an underpinning, however. We can whimper all we like about the fundamental illegitimacy of the state, but without a systematic inquiry into the philosophical foundations of justifications for the exercise of power, all of our words will be pointless and inconsequential. The vital thing, if any complaint about the government is to be taken seriously, is for it to be grounded in a carefully considered social contract theory. We must know whether the relationship of state to citizen is akin to that of father to child, mother to father, or son to lover.

Every multilayered philosophical lasagna must have a bottom onion, and where social contract theory is concerned, this is the notion of *tacit consent*. If we may determine the boundaries between that which we (as people) have willingly subjected ourselves to, and that which we have not, at last we have the muffin of ontology that can be icinged into the cupcake of general ethical theory.

We consent to many things explicitly. For example, when a human signs a contract for the purchase of a house, she has agreed to its terms explicitly. In giving it the old "Jane Hancock," she has said "Within these walls I will place a bed, and on that bed I will sleep, and in exchange I offer a thing, and it is called money, but might as well not be, and that is as it should be." And her Realtor has replied, "Yes, exactly."

We accede further to many things implicitly. An example in this category is our consent to the way in which the meat that we choose to buy is produced (by choosing to buy a certain sausage we are supporting the activities of the manufacturer, however ignoble or humorous.) By hanging up my coat, I tell the barkeep I am a tidy man. But where do we draw the line as to what we tacitly consent to and what we do not? What am I responsible for? What have I agreed to?

But perhaps there is no line to be drawn in the sand or even on the nearby promenade. Perhaps instead of delineating the big clear sphere of moral responsibility, we ought simply to accept a generalized mushy principle that we tacitly consent to all actions taken by those groups with which we associate, and ought to increase our sense of responsibility for those activities conducted in the name of our nation, religion, or housing unit. Perhaps instead of elaborating a complex theory of political obligation as a rationalization for our infinite daily shirking, we must exist as ourselves in a state of perpetual protest against that which we do not wish to consent to.

No doubt, malicious book reviewers here will pounce, and insist that a robust theory of moral responsibility for that which we never even agreed to in the first place is at best, impenetrable, and at worst, impossible. If I cannot stop that which the United States or the Aryan Football League does in my name, how can I be held responsible when either of them torpedo an elementary school and tell the world it is because they love me? But the reviewers have forgotten the possibility of suicide. For it is always possible to revoke one's consent. We may remove ourselves from the roster, so to speak, at any time we please. And thus so long as this possibility exists, we continue to tacitly consent to each presidential assassination attempt conducted by our infatuated fan club secretary.

As human beings, then, our political responsibilities are clear. We must consent to tacit consent, *kill ourselves, or dismantle all of our social relationships until we are fully aware of all of the obligations we have.*

DEATH TO SOCRATES

But as we have momentarily taken port on the island of Political Theory, let us use the opportunity to digress on one of its most over-
praised salesmen, pausing to correct a serious historical error in philosophical character appraisal.

Socrates, that infernal corrupting bloviator, has been afforded
a place in the West akin to that of Dickens or Cousteau. Yet what
are the crowning achievements of this hirsute Athenian gasbag? Has
he penned a memorably perceptive travelogue or snorkeled a hither-
to-unsnorkled lagoon? For a career that consisted in the main part of
playing the tiresome five-year-old shouter of "Why?", Socrates has
been treated as a martyr for freedom, instead of for irritation.

The central problem with Socrates is *not* that he is a sponging
bearded malcontent. It is, rather, that his celebrity allowed for the subsequent veneration of the Socratic as a dialectical ideal to be aspired
to, rather than a nadir at which we should ralph. Consider a represen-
tative snipping from the Dialogues:

> **Socrates:** Would you agree that a two-sided thing can never
> be square?
> **Mamelon:** Of course, Socrates.
> **Soc.:** And that a square is even in its sides?
> **Mam.:** Yes, Socrates.
> **Soc.:** And the even is the orderly,
> and the orderly the even?
> **Mam.:** There can be no doubt.
> **Soc.:** And the state acts as a monolith?
> **Mam.:** It does.
> **Soc.:** It would be absurd to speak of a "two-sided" state?
> **Mam.:** The Gods would scoff, Socrates.
> **Soc.:** Then, dear Mamelon, if the square is perfect order, and
> the square cannot be two-sided, and a two-sided state would
> likewise be absurd, then perfect harmonious order can only
> be found in a powerful centralized state. One ruled by phi-
> losophers.
> **Mam.:** Why, it could not be otherwise, Socrates!
> **Soc.:** It is elementary.[44]

---

44 from Plato, *The Geometries, Book IV* (Jowett, trans.); In our analysis and formulation of
the bearded one's treacherous unreason, we crib liberally from Émile Faguet, *Pour qu'on
lise Platon.*

Any child educated in the prating sycophancy of the Socratic "dialogues" (actually little more than monologues interspersed with verbal applause) may not only develop worrying inclinations toward pederasty and metaphysics, but will be rendered incapable of performing a social function other than as a servicer of government and corporate power comparable to the storied "fluffer" of the pornographic film set.

Are there redemptive features nevertheless to be found deep within the gristle of the Socratic paunch? No. The man's last words, *"Aloha, ma belle monde!"* may have become a bestselling catchphrase for the novelty-mug-and-magnet racket (perhaps exceeded only in ubiquity by "My Other Dad is a #1 Grandpa"), but this says little; even the Chiquita corporation could come up with an enduring prophetic jingle. No number of vials of philosophical antivenin can counteract the coursing paralytic toxin of Socratic thought and discourse.

It is time, then, to beseech the academy to cast its favourite founder into the volcano. The salivating jaws of the dustbin of history await tender new refuse. Go and fetch the hemlock.

If Every Dollar Is A Vote, Then I Am A Serf

But life is too short to spend any of it studying history. Let say goodbye to Plato and the man he lived vicariously through, and turn from political philosophy to monetary philosophy. In sharp contrast with the Marxist tradition, we believe that an understanding of economics is important for the making of political pronouncements. After all, many of us interact with economics in our day to day lives. Further, we have a long history of endorsing cross-disciplinary pollination, a belief one of us acted upon directly when he married an ethnomusicologist who had published *The Annotated Tupac*, an affordable volume directed at a non-specialist audience. And though said ethnomusicologist turned out to have just as strong a scholarly interest in cavorting shamelessly with her two-faced periodontist as in theorizing the rhythmic diaspora of Afro-Caribbean syncopation under post-colonialism, this does not in itself prove that investigating topics outside one's subfield is unwise.

It has become common to assume that money does not matter. *That's all very well, but money can't buy me love*, shout the tiresome

Liverpudlians as they incessantly revolve. "Yes," say we, "but it can get you a steak dinner, and one cannot love without being well-fed." And in one sense, money not only *does* matter, but *is made of* matter.

"What concerns it me if my neighbour spends a billion dollars purchasing a mile-high dolphin enclosure to placate his heartless step-son?" is the usual reply. The assumption is that *so long as I have mine*, I need not concern myself with what absurdities others are brewing on their own turfs. It is not inequality that matters, but individual well-being. If I am being well, the number of damns I should give about rich men's follies approaches zero.

But this is a miscalculation of proper damn-ratios. For it ignores the elementary mathematical truth that *every dollar a person who is not me has is a dollar that is not had by me*. Every time the barber next door acquires another dolphin, he deprives me of an opportunity to mobilize the same social resources to far more reasonable ends. We could have taken the entire neighborhood on a Mediterranean holiday, instead of simply furthering one man's brooding step-teen with yet another opportunity to gawp momentarily at unnecessary sealife under the mistaken theory that spoiling the boy will cause him to embrace and adore his mother's pitiable new spouse.

With money, one may at last do as one pleases. When one's sanctimonious Dean informs one that "professorial transvestism" distracts from the content of one's lecture, if one is comfortably well-off, one may simply tell him to go and boil his head (or worse). If, however, one is surviving on the pitiful salary of a co-assistant lecturer, one may be forced against one's better judgment to remove one's sequined corset.

Money is not a government, they say. But a glance at the sentence reveals it to be untrue. For while it is indeed the case that money does not standardize or operate its own postal service, money is nevertheless an unforgiving truncheon not unakin to the policemen's. Money may not close for Columbus Day, *but that does not mean it is not a bank*.

Consider each dollar as being akin to a vote, a vote over how a particular small fragment of the sum total available human resources are to be used. Thus construed, inequities in financial resources become far more worth gaping one's mouth or tilting one's eyebrow at. If the mayor were to arrive at our door and say "From now on, Mr. Robinson, you shall have four votes while the man who thought iPads were a good idea will have forty billion votes," we would be so shocked as to draft a series of infuriated letters-to-the-editor. It is fundamental that all humans ought to receive the same number of votes (unless they happen to be resident aliens, in which case their right to decisions over their political representation is universally agreed not to exist.) Yet since money operates precisely the same way votes do, how do we thus justify not allowing each human precisely the same quantity of funds? With money construed as little more than a queer form of ballot, the necessity of full communism is made irrefutable. If every dollar is a vote, then I am a serf.

When Everyone is a King, Everyone is Still a King

> "One day an official from the palace arrived with a dead dolphin in the back of a truck. He said the leader wanted it stuffed. The museum staff protested that this was impossible because a dolphin's skin contains too much oil. Mahmoud laughed as he remembered the terrified expression on the official's face when told that Saddam's order could not be obeyed." [45]

Marcus Garvey died amusingly, but surprisingly enough he did not drown in one of his enormous hats. In fact, he had a stroke.

Garvey was incapacitated by an initial stroke in January of 1940. George Padmore, who was a columnist for the *Chicago Defender*, had heard a rumor that Garvey had died. Instead of confirming this, Padmore published a premature obituary. The obituary described Garvey as a man whose followers had deserted him, a man who died "broke, alone and unpopular." When Garvey read it, he let out a loud moan and he collapsed. He suffered a second stroke and died the next morning.

The moral of the story will be clear. Garvey aspired to regency and found nothing but contempt. He who seeks the throne is most destined to be spattered in the sputum of humiliation. Autocracy is the drug of the accidental narcissist.

But occasionally proposals are issued that nominally undo the problem. "Yes, yes, nobody likes a flaming despot," says the eager

45   Patrick Cockburn, *The Occupation*, p. 77.

liberal. "But the solution is not to eliminate despotism, but democratize it. Give each a patch of turf over which she can reign supreme, zapping ants, sculpting lewd topiaries, etc. The problem solves itself."

Yet the solution to hierarchy is not hierarchy. To democratize tyranny is to make a million tiny tyrannies, each governed by a potentate the size of a garden fly. It leads to the proliferation of municipal busybodies on miniature power trips over their small sector of local administrative regulation. It is no accident that the BTK killer turned out to have been a dog catcher all along.[46]

This runs somewhat contrary to positions we have previously announced. We have, for example, stated strongly in one of our more popular works that "when everyone is a pope, no one is."[47] But we used this phrase only in the hopes of selling books; you will forgive us, but there was a time in the recent history of literary nonfiction when agents were accepting query letters only if the works in question somehow incorporated favorable comments on the papacy.

It is not true, then, that handing power out like free supermarket marinara samples will result in the total liberation of mankind. When everyone is a king, *everyone is still a king*.

RESPONSIBLE MONARCHY

Yet let us consider whether there is some salty redemptive kernel popping amid the global monarchy. Because kings have largely been portly and bothersome, there is a tendency to become biased against them. "Why should Leopold loll about castles and luxuriate in flounced robes of human skin when I work in a gift shop?" asks the peasant. It does not seem *fair* to have a king when one cannot oneself aspire to the position, which no number of fabrications upon one's *curriculum vitae* can help one to obtain.[48]

But to throw every king into his own volcano would require

---

[46]   Gary Younge, "Dog catcher admits serial killings," *The Guardian*, June 27th, 2005.

[47]   Oren Nimni, Nathan J. Robinson, & Orval Faubus, *The Decentralization of Dogma: A Populist Popery for the 21st Century* (Opaque Theology Press, 2007).

[48]   We refer here (obliquely) to the classic Mexican folktale about the young boy Pepito, who sat about embellishing his résumé until the day he was finally eaten by a tiger. The lesson is sobering. See "Pepito and the Saguaro," in *Cuentos Populares de la Madre México [Popular Folktales of Mother Mehico]*, Smythsonyan Institution Press (Unaffiliated With Namesake), Nimni-Robinson (eds.), p. 740.

numerous volcanoes. Perhaps the problem is not monarchy, but monarchists. After all, kings themselves would largely be content to suck on marzipan and occasionally go out to inspect the troops. Their flacks, hacks, lackeys, flunkies, and lankies, however, do incalculable damage to the national *bien estar*. For it is not the king himself who organizes the parade in his honour; it is the king's Royal Chief Paradesman. The problem with nude emperors is the conspiracy of the village to mentally clothe them, *not the fact that they are nude*. Pageantry can be a gas, even a noble one; the problem is that monarchy goes to excess. We worship our kings instead of occasionally seeing them out buying vegetables and stopping to say hello. Consider this notorious double-sentence from the *Autobiography of Henry VIII:*

> *Nid oes gennych unrhyw syniad sut syniad pa mor fodlon y mae'n gwneud i mi i lofruddio fy ail wraig. I ddinistrio elynion un gyda gael eu cosbi; mae hyn yn wir wefr o frenhiniaeth.*[49]

Not exactly the charming Cockney rascal that the pop songs have portrayed him as. But the point remains: kings must have their scepters snapped and their orbs castrated if they are to serve their redemptive theatrical function without also causing a massive imperialist slaughter or blowing the treasury on a thousand-acre hedge maze.

We can picture a responsible monarchy, however. A queen or king much like the American president: impotent and ceremonial. Strip the leader's functions, let him wander about in costume and exhibit his attractive offspring. Let him shake whatever hands he pleases without said shakes constituting binding international tariff agreements. Let queens and presidents and prime ministers all take on the same identity: a charming individual who gives merry waves and boosts magazine circulation with occasional pregnancies. But for goodness' sakes, let the reins of power be kept carefully locked away in the weapons pantry, never to be entrusted to any such people.

Yet in many ways this discussion is moot. After all, there is a king even more ludicrous and tentacled than any Charles or George, one who has lately become the Central Animating Precept of human institution-building. We speak, of course, of the bureaucracy.

---

49 Trans.: "You have no idea how happy it made me to murder my second wife. To destroy one's enemies with impunity; this is the true thrill of monarchy." Any and all Henry VIII quotations in this work are taken from the 1648 edition of the memoirs, which most now agree to be a transparent forgery.

BLUES AND THE BUREAUCRACY

The twentieth century is popularly conceived of as a vast struggle between the deep C of communism and the high C of capitalism.

But we reject this paradigm with all deliberate gusto, and instead propose a new grand dichotomy for contemporary historical analysis of the preceding 100 years. That which drove all social relations and all political machination was in fact the following: the unacknowledged war between the Blues and the Bureaucracy.

It is tempting, certainly, to reject our thesis on the not-unreasonable grounds that C follows from B, and therefore represents a *progression* of thought. The Blues/Bureaucracy conceptualization, it is argued, marks a *regression* of contemporary analysis. But this rebuttal is so deeply flawed as to be incorrect. For, by the same logic, a paradigm which saw the conflict of the 20th century as the conflict between Anthropomorphism and Arthropods would represent the basest and most primitive mode of thought, when we are all aware that this is not the case, due to the superior numbers of syllables present within the words in question.

What, then, is blues? And what is this bureaucratic counterpart it supposedly carries around with it? Perhaps Ms. Bessie Smith can shed a bit of light:

> *Gee, but it's hard to love someone when that someone don't love you*
> *I'm so disgusted, heartbroken, too*
> *I've got those downhearted blues*
> *Once I was crazy 'bout a man*
> *He mistreated me all the time*
> *The next man I get he's got to promise to be mine, all mine*

In Ms. Smith's telling, the blues sounds significantly miserable. How, then, can it stand opposed to bureaucracy, which is also a creature of misery? Part of the answer lies in the question. Bureaucrats and bluesmen are opposites because B1 *causes* B2. Another part of the answer, however, is more complicated. Blues may be musically-embodied misery, but blues simultaneously engenders *relief from misery* through the expression of it. Bureaucracy, on the other hand, simply makes *everyone* unhappy, bringing scant relief and few danceable grooves. Blues may *seem* downtrodden and resigned, but it is also liberatory, in that it gives a poetic outlet to feeling rather than allowing it to fester and decay within the bluesman's potable innards. The bureaucrat clads himself in grey flannel and finds no method of sublimating the downward pressures of the institution, thereby eliminating not only his sexual capacity but also his personhood. The bluesman, by contrast, is both sexual and a person.

This topic naturally requires full exposition over the course of a multi-volume scholarly treatise. We offer this summary of the basic dichotomy in order to encourage young academics to pursue studies in blues/bureaucracy revisionist history. There is, after all, a reason B.B. King's initials were as they were.

OTHER BUREAUCRATIC MOLASSESSES

But let us go into more depth, and cover ourselves thoroughly in bureaucratic molasses, while nevertheless accepting the precaution that too much molasses can indeed be a very good thing.[50]

Bureaucracy is actually a mixed pestilence. On the one hand, nobody wishes to sit in the waiting chamber for four hours only to have the customs officer deny her auctioneer's license. On the other hand, without any formal procedures at all things can become slightly chaotic. We well remember what happened when the negligent incompetence of a wayward undergraduate research assistant caused us to end up facing a vindictive departmental witchtrial on trumped-up plagiarism charges. In this (quite literal) kangaroo court, we shouted in vain that *proper procedure* necessitated our being allowed to subpoena the student's parents to testify regarding his history of shoddy researchmanship. But the dean, despite being a truffle-snuffling sow when it comes to unearthing academic cribbing, is an anything-goes anarchist when it comes to enforcing ordinary rules of civil legal procedure in a university disciplinary hearing. We were therefore unjustly prohibited from introducing the key evidence, on grounds that doing so would necessitate enacting the *unreasonable*.

But "be reasonable" is not and could never be the cry of the

---

50   In saying this, we do not wish to offend descendents of the victims of the Boston Molasses Disaster, a crime of capitalism for which the nation has never sufficiently atoned. See Stephen Puleo, *Dark Tide: The Boston Molasses Flood of 1919* (Beacon Press, 2004).

bureaucrat, who knows that where rules and reason cross streams, it is ever reason that must get wet. The bureaucrat would have had no qualms granting our subpoena, indeed *could not have imagined it otherwise*, for to him the rules must be enforced, however hilarious the consequences.

Anyone who charges that we are anti-bureaucratic is therefore worryingly disconnected from the historical record. After all, were we not the ones who, when a certain member of the Political Science Department parked her Volkswagen in a space reserved for members of the Sociology Department, insisted that the little wizened campus towing man carry her Vanagon off to the crusher? We were indeed. And would those who prioritized common sense over bureaucratic sticklerism conduct such an insistence at such a volume? They would not.

But even though, then, we enjoy bureaucracy when its procedural ejaculate gums up our enemies' embossing machines, we nevertheless think human beings largely deserve better than to spend their eternities gooily tarred in the thick morass of bureaucratic molass. The humble aesthetics of dingy brokerage halls and broken water fountains may carry their unique charms, but one can only fill out one's name so many times before one begins to sense a cosmic tedium incompatible with the perfectly just society. Nobody whose grandmother has been euthanized at the border because he failed to properly fill out Part B(a) of his Grandparental Pro-Vivos Declaration can possibly be unreseveredly pro-bureaucratic. And as those who have experienced this aggravating misfortune not once, but eight times between the two of us, we come to the hearty if reluctant conclusion that *the bureaucracy must be done away with*, preferably with some haste.

Probably, Actually, the State Itself

Yet perhaps we should encircle a larger sphere than mere bureaucracy, and consider the question of whether the triumph of reason requires destroying the state itself.[51] After all, *the State is as the Octopus*.[52] And octopi, as any first-year marine biology major learns on page one of *The World of Contemporary Sealife*, are too powerful to be allowed to live.[53]

Now, when we find ourselves advocating destroying the state, we experience a bit of a dilemma. For, on the one hand, states across the universe have historically done little more than at worst oppress and at best inconvenience. On the other hand, some problems require giving the occasional evildoer the occasional boot to the chest, and often only the State keeps large enough boots handy.

One should not rely on the state to force itself out of existence. If we learned one thing during our brief engagement as folktale harvesters on the Works Progress Administration's dime (and during the subsequent expense-account audit hearings), it was that one cannot legislate violence, for violence is inherently absolute and legislation gets watered down in committee by cowards.

The position we would announce, then, is a tentative and cautious one: the state must be destroyed, but not carelessly. To collapse into chaos and interminable working groups is undesirable, but so too are death camps and sex drones. If we need a state to make a Tennessee Valley Authority, so be it. But let us not thereby be deluded into believing that life itself is a T.V.A.

---

51   *"The secessionist impulse is the human impulse, imbued within us at birth."* - Aphorism #10.

52   Aphorism #11.

53   Obviously, the crimes of the state need no elaboration here, and will be familiar to all those for whom noses and brains (arses, elbows) can be distinguished.

III. ORTHODOXIES

Well, now, doctors agree
So I've been told
Do the twist and
You'll never grow old
- Gary "U.S." Bonds, "Dear Lady Twist,"
Legrand Records, 1961.

Our revulsion at the more noisome aspects of the present-state does not preclude our recognition of its unique pleasures. Even the most artistic of historical peoples tend to have some sort of minor impressive innovation performing a small jig upon their resumes. The Chinese invented the stirrup, for instance.

All our yesterdays have not, therefore, been comprised solely of foolishness and dusty death. One or two of them managed to spew forth a lasting piece of human worth, such as The Twist, or at least managed to impregnate a future generation with a speck of possibility, in violation of admittedly laxly-enforced prohibitions on inter-generational copulation.

It is not historically sufficient, then, to stand on the rooftop of the present shouting "Death to all!" Such an act would convict one of that gravest of capital offenses, *lack of nuance*. "Go and be a fondler or a forger, it is no business of mine," says Respectability. "But nuance, dear boy, never lose your *nuance* lest I flee your side evermore."

In this section, were therefore reluctantly catalog the things about our present age that are *actually okay*.

A. Structure and Infrastructure

REVIVING THE "UNIT"
WHILE DISPOSING OF THE UNITARY

The use of the word "unit" in an earlier chapter is by no means unintentional. (We do not take kindly to being accused of co-incidence.) As can be discerned from the division of our book into chapters, the unit is the central component of our thought.

Yet before fully embarking on Part III, Section A's investigation of redemptive orthodox structures, we would be remiss if we did not note a certain creeping slovenliness in the modern intellectual approach to the unit. This can be summarized as: *the devolution of the unit into the unitary.*

The difference between the two is fundamental. "Unit"-based modes of thinking see man and his surroundings as an organic system, into which stimuli are placed and from which responses are drawn. A unitary view of nature and man, by contrast, sees the individual not as operating in concordance or conjunction with other elements, but independently of all outside factors. The confusion, then, is inexcusable.

The unitary view of things, a clear whole rather than a whole of parts, has repeatedly been shown to be out of touch with really existing conditions. Yet there are those among our readers who will place no stock in the scientific consensus. For these provincials, we present the words of an actual, flesh-comprised President.

"The willfully idle man, like the willfully barren woman, has no place in a sane, healthy, and vigorous community... Exactly as infinitely the happiest woman is she who has borne and brought up many healthy children, so infinitely the happiest man is he who has toiled hard and successfully in his life-work."

- Theodore Roosevelt,
"Speech at the Minnesota State Fair,"
September 1, 1901.

Roosevelt spoke in metaphor, but he spoke clearly. It is the unit, he said, with which we must deal.

However, this is only the beginning of the task. While there may be universal agreement upon the affirmation of the *need* for positive and resolute living environments, and the machines/vehicles contained within, such common understandings fail to address the core dilemma: *How will these living environments be divided, and will humans exist as atomized specks or mutually-dependent co-habiting organic units?* The stuff of life may be *rich*, but unless it is divided according to the principles of an informed manifesto or scientifically-sound philosophical foundation, we will find ourselves wandering through life with a perpetual sense of vague disquiet. We may decide to divide, or not to divide, but either way we must decide. Thus, throughout this Part, *keep in mind the theme of the unit* as being the main intellectual tarpaulin atop the disused mower of our notions.

THE FRONT PORCH

We do not often ask the question "What ought to be in front?" for the meaning of the inquiry is less than obvious. But in the case of housing, the answer has ever and always been self-evident. Has any house without a front porch ever been better than any house with one? Never.

This is because *architectural determinism* reaches its apogee in the porch. Without a porch, we do not get to sit sipping whiskies and waving hello to our neighbors; with one, we do. Thus, the porch *inherently* determines the level of community-spiritedness a particular city block or country acre experiences. We are turned amicable by the porch; the porch befriends us to the world.

Our own advocacy of front porches has even extended as far as the charitable. We have donated porches to the poor inhabitants of bleak modernist condominiums, and have built them around the Dean's stately home as he slept to surprise him in the morning. Those, like the Dean, who see the application of porches to a problem as grounds for indignation and the pressing of charges, are beyond help.

The porch is a precondition of progress. Lest we sit upon it in rocking chairs, we stray from the righteous course.

The Parade as Centering Device

But the porch only assists the house itself to be neighborly. It is not *in itself* a *unifying act*. An act and a structure are quite different entities indeed.

Do you remember when you were happiest? It was at the parade. For it is at the parade where human individualities and distinctions are at their most erased, and we each become afloat.

One questions the parade's usefulness at one's peril. The quantities of egg that parade-rainers have found slathered across their faces more than exceed the annual production capacity of the nation's brutal factory farming industry. For parades are very, very enjoyable, and there is a reason each of us remembers little about our childhoods except the parades we attended.

A town has a tendency to become decentered. People wander off into various careers, split into their various divorces. A high school graduating class of 100 might produce hairdressers, lye salesmen, and telecom lawyers, all without changing the curriculum! But when these disparate citizens need corralling, in what does the town find its lasso? *In the big parade.*

It is easy enough to argue about what love is.[54] It is less difficult to dispute what a parade consists of. Gather the unions, marching bands, leftover veterans, and anyone with an amusing automobile; line them

---

54 As our own tender coinage goes, *"Love rots ephemeral, but parmesan springs forever."* (Aphorism #12.) In actual fact, this is not strictly an aphorism *as such*, but rather an advertising slogan conjured during our brief tenure as scholastic marketing consultants for the Pacific Northwest Dairy Board, who wished to expand the use of the "secondary" or "incidental" cheeses among young people. It demonstrates the lasting wisdom given us as a result of years of crushing heartbreaks as well as numerous surprising discoveries that, amidst the ruins of former romances, a tube of parmesan thought to surely be long-expired was still as fresh as the day is unnecessary.

up and send them forth. A billow of mass happiness envelops the onlookers, and the civic spirit survives to stagger through another day.

Parades, then, will never stale. They may unwind, go haywire, or lose the route and plummet into a manhole. A kitten may be trampled here and there. But at their very best, they exemplify all that is most procession-oriented about America. And even at their worst, they are nothing more than a massive time-wasting disappointment.

THE PARKING GARAGE AS LATTER-DAY BIRTH CANAL

It is said that a picture is worth a thousand words, but the picture below is merely worth the entire net positive contribution of this condition known as mankind from its origins to the present.

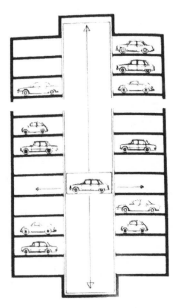

The parking crisis is analogous to the reproductive crisis, in that each concerns an entity which continues to grow until it is either transformed or collapses. If the female of the species allows her progeny to fester inside her womb past the birthing age, she will find herself unsustainable. Similarly, as the use of motorcars continues, we must either innovate upwards or face the prospect of shattered reproductive parts.

As illustrated by the leftward diagram, bi-axled motorized transit has thus far followed a North-South-East-West model of parking inquest. When a car wishes to park, it may move in any direction, so long as this direction fits within the two-dimensional schematic we offer to the left. But as its use grows, the automobile finds itself impotent, incapable of moving without creating either oppressive traffic conditions or plunging into the sea. For the North-South-East-West model is premised upon a delusion; namely, that there is room for infinite expansion as far as parking is concerned. But as the CEOs of America's largest shopping-mall complexes can tell any of us, this is so far from the case as to not be the case.

Catharsis through parking might be a theme of "socially-con-scious" literature,[55] but it cannot hope to sustain itself over the long term. Not all minerals can become cars, yet our population and car-per-person rates continue along an inverse plummet. It may be true that we see emergence from the darkness of a particularly enormous parking garage as a kind of rebirth, yet we cannot allow our desperate need for simulations of a return to the womb to justify the enormous psychological and environmental toll exacted daily by the automobile.

What it comes down to is this: If most automobiles are station-ary all of the time, surely we could eliminate most of them through a system of communal parkingness and borrowance. In the diagram above, all of the vehicles are still, and many parking spaces are empty, yet both objects nevertheless exist. The maximization of efficiency so treasured by the bowtied economic-types demands that all existing cars be kept in motion at all time, regardless of the effect this may have on birth-canal fantasies (which cannot be graphed, and therefore must be discounted.)

THE POET AS ARCHITECT

The parking garage shows how structure and the life that occurs with-in are precisely correlated. Likewise, the correspondence between equations and poetry has been vastly overlooked by scholars of both English and mathematics. While there have been occasional literary attempts to blur the fields, such as Edwin Abbott's legendary geom-etry thriller *Flatland*, or George Orwell's famous numbers drama *One Nine Eight Four*, they have missed the fundamental link between the two arts which we now propose to elucidate.

The principle is best demonstrated through example. Below is a simple blueprint for a modern-day motorcar, of the type popular among both adults and young people:

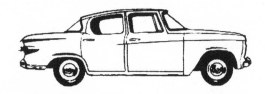

---

55   We have gathered that this is the theme of a novel by Joni Mitchell exploring the idea of the parking lot as paved paradise.

We have attempted to utilize the most basic of the motorcar-designs available, for the purposes of illustrating the poetic-geometric principle with the greatest clarity.

Now, here is the first stanza of the poem "In the Garden at Swainston" by Alfred, Lord Tennyson (1809-1892):

> *NIGHTINGALES warbled without,*
> *Within was weeping for thee:*
> *Shadows of three dead men*
> *Walk'd in the walks with me:*
> *Shadows of three dead men,*
> *and thou wast one of the three.*

Astute readers will immediately note the similarities between the rounded cap of the automobile's body and the rounded "B" of the word "warbled," which serves as a similar ceiling to the poem. But this is only the tippy-top on an entire iceberg of similarity, which stretches from the windshield-words of "thee" and "men" to the wheel wells of "three" and "thou."

Because this is a popular and not an academic text, we will not here include the mathematical formula for the derivation of geometric truths from poetic ones. We will, however, caution the reader that no attempt to pilot a motorcar can be undertaken without a corresponding attempt to navigate the heart of a work by a 19th Century poet.

THE ARCHITECT AS POET

But if, as we have previously established, each word is a geometric line, any attempt to "draw" or "sketch" crafts a "poem" at the same time as a representation of the object it seeks to represent.

This is a principle which found a safe home in the writings of Christian architectural socialists like John Ruskin, and yet it has been given short shrift by those who currently work in either of the fields in question. We attribute this to cultural decay and to the corrosive tendency of the modern university student to hang around massage parlors and olive groves.

As a corrective measure, we suggest the reintroduction of the whip to university classrooms. Flogging has been proven a satisfactory solution to the better half of human problems.

B. Society, Culture, & Animals

HARNESSING THE POWER OF THE SCIENCES

So answer me, illustrious philosophers, those of you thanks to whom we know in what proportions bodies attract each other in a vacuum, what are, in the planetary orbits, the ratios of the areas gone through in equal times, what curves have conjugate points, points of inflection and cusps, how man sees everything in God, how the soul and the body work together without communication, just as two clocks do, what stars could be inhabited, which insects reproduce in an extraordinary way, answer me, I say, you from whom we have received so much sublime knowledge, if you had never taught us anything about these things, would we have been less numerous, less well governed, less formidable, less thriving, or more perverse?

- Mr. Jean-Jacques Rousseau,
Discourse on the Arts and Sciences

We simultaneously reject and embrace Mr. Rousseau's conclusion. We are in firm agreement that the armies of lab-coated test-tubers that roam our lands have done little to nothing to break the shackles that bind us. If we had our way, all persons would inhabit the Enchanted Wood, where they could be free of nuclear waste and cotton gins and steamships and all of the other so-called "labor-saving" devices we have been given thanks to their munificent benevolence.

And yet we do not share his resignation or distinctively French crotchetiness. There is hope for a liberatory technology! Why can our pants not be folded and our groceries totaled by Machines rather than Man? What is keeping us from converting ourselves into the hedonistic beneficiaries of indentured robot-labor? Our dignity? Our maddening insistence on job retention?[56] Our inner fear of decapitation at the hands of our mechanical butler? Fie on such concerns. We, for one, will be first in line when the Intel Corporation unveils its first self-playing piano or automated tea-pourer.[57] Anything that relieves the tedio-drudgery of the Working Life is to be irrationally and enthusiastically adopted.

It is true that many of our Modern Marvels moonlight as psychotic killers. It is true that our Internets and Televisual Entertainments have hastily zombified the populace and numbed us to the Injustices. But as we toss the soupy contents of our used bathing tubs into the street, let us ensure that there are not unseen babies lurking within. If technology cannot be transformed, then away with it! But this is a position of extremity, and experience suggests that moderation is a more appropriate state of approach in the circumstance. Why, just look at the Bananaphone! Has it not given ease and comfort to those who previously went without? Has it not revolutionized our telecommunicative and nutritional interactions? Those who say it has not have clearly misunderstood or misused its wonders.

Nobody is more prone to shivering with unwarranted terror at the State of Things than we blueprinters, yet we cannot fully join Mr. Rousseau in his condemnations. We hold out hope that a carefully-guarded technology can bring creations of untold majesty, and that we must not sacrifice the eternal dream of Bean-Powered Jetpacks or Waterslides From Space for the sake of allaying a base trepidation.

But *is a moral science possible?* And by this, do we mean a science *of* morals or a science which *is* moral? Either meaning could be gleaned by the crafty gleaner. Our proposal, in the spirit of the Great Concil-

---

56   Along this line: from what source does the State of New Jersey think it obtains the right to prohibit two passing-through professors from pumping their own gas into their own '93 Sunfire? The legal mandating of useless labor is a crime against even uselessness herself. Nevermind the edible butler, what is the point of paying persons to pump instead of paying them to build vast crystal domes and letting every harmless well-educated freeway motorist refill his conveyance without encountering illegitimate and purposeless bureaucratic meddlesomeness?

57   As the old saying goes, "You can tuna piano, but not without professional tools and years of intensive training.." (Aphorism #13.)

liators of the age, involves a fusion of both concepts: a *moral* science of morals. In this fashion, we can both derive moral maxims through Reason and Experimentation *and* pride ourselves on the morality of this very deduction, leaving no room for accusations of shabby nihilism or disregard for Life.

Science is not pinochle. Yet can a trade which sustains itself on H-bombs and vaccinations ever be considered truly noble? Can a field in which the sole criterion for measurement of success is the number of smithereens produced redeem itself in the eyes of the Public? Is it even possible to force morality to conform to a positivist aesthetic? If Kissinger is taken to be the foremost "life-scientist" of the age, the answer is necessarily an emphatic "No." Remember too, the horror wrought by George Washington Carver, inventor of the peanut allergy.

But if we have science, we must have morals, lest we become the very test subjects we ultimately hope to exploit. A scientific morality cannot easily be cultivated, however, as science tends to involve objects that tend to exist, while morals tend to be pulled from the collective rectum.

If scientists quit building velociraptors and began building bookshelves, would we even be in this mess? Would we have wreckaged our earth-mother and disparaged the seas with fishwaste? Perhaps. Perhaps it was the destiny of humankind to rape its blessings and then destroy itself. But we are optimistic that a new kind of pessimism is possible.

CURES FOR DISEASES

Every time some precocious young preteen pant-sagger comes up to us on the bus and shouts some hip catchphrase about the crimes and uselessnesses of modern science, we find ourselves propelled into a cloud of reflection. What is it that science has done for us, precisely? Did it prevent Mama Rouge from an early death at the hands of the Tetanus Monster? It most assuredly did not. Has it put a turkey in every pot? Not in either of ours, although it did cause a heinous infestation of wild turkeys at the Nimni household one spring, who refused to depart until they had gobbled through each and every item in the prized Nimni Collection of signature frontispieces.

It must be said, though, that for every disease science has failed to cure, equally has it cured one. Alongside the persistence of the measle came the eradication of polos. Would those poor wretches subject to instantaneous development of the "Roosevelt disease" (as Grandmama Nimni used to dub it out of a distaste for aristocracy) for a moment burn their Science Draft cards or Ralph Lauren handbags knowing how it all turned out? Even the painfully bemeaseled would hardly begrudge others their vaccinations. Give a man a fish and he'll eat it, give a man the measles and he'll complain vociferously. Cures and medications, then, may well be a redemptive product of science.

Remember, however, that cures are the dyslexic's curse. There are some diseases for which the cure is actually an ointment. Who among us would dare to cure brain-death, for instance? The caveat would do well to be borne in mind.

A final quotation should sufficiently clarify the position:

> *What must one think of [Cézanne,] who spent all his life trying to paint round apples, and who never succeeded in painting anything but convex apples?... One has to be extremely awkward to be content with painting apples that are such a failure that they cannot even be eaten... Le Corbusier also made a disgraceful mistake: never will reinforced cement be used on other planets...Yes and yes, he sank like a stone, the weight of his own reinforced cement pulling him down like a masochistic Protestant Swiss cheese. On a structural basis, Cézanne is like Le Corbusier; the only difference between them is that Cézanne was a rabid reactionary and full of good intentions whereas Le Corbusier was irremediably Swiss, left wing, and full of bad intentions...Buckminster Fuller has freed architecture from the right angle and has substituted for structures that are heavy others that seem to take flight; he has demonstrated that the ideal shelter for man is a spherical translucid structure which might cover the earth—a cupola!*

- Salvador Dalí, from <u>Dalí by Dalí</u>, 1970, pp. 40-42.

Dalí makes clear that if we are to speak of scientific and medical progress, we must also talk of technology itself. For what is a science without the innovations it produces? The question must then be asked: how are our everyday lives conditioned by the devices which are created under conditions of rapid-scale progress?

THE ROBOT BUTLER

Each of us serves only ourselves, but certainly a butler can be a useful supplement. Naturally, our own staunch left-wing politics prohibit us from employing butlers, but we have thus far managed to deploy graduate students toward similar tasks for a fraction of the expense.

But isn't the dream to have one's butler and eliminate it, too? Or, to put it another way, to eat the butler without having him be eaten? Nobody consciously wishes to buttle; nobody sets out down the misbegotten garden-path of a service industry career with an intention to serve industriously. That is to say, *infants* do not dream of someday becoming *servants*. And yet, were it not for servants, would life not be slightly more difficult to live? It might indeed.

Thus a key problem facing the innovator and the theorist is the necessity for service without servitude. We all wish to be given a rental car or a bucket of french fries without ourselves having to operate a deep fryer or oversee an auto plant. At the same time, the existence of fry-slaves and car-people is a moral horror. Why should your ordinary Joe Fourbiscuit have to spend his life shouting "Order 98 is ready" into the cosmic void, *especially if Order 98 is not yet ready*? Do we accept as necessary the fact that certain individuals may spend decades collecting spinal injuries from plucking tomatoes and rotating hospital beds, while others may pass weeks at a time in a bathrobe jotting unreadable monographs on motifs of the mule in ancient Hebrew folktales? Is there a word for this moral chaos other than sordid?

But it is here that the Robot Butler enters through the French window with a jaunty "Yes, sir?" For though robots are people, they are nevertheless not humans. A robot does not mind which way one roughhouses him. One can stroke his circuits giddily without committing an act of turpitude. A robot is a friend, but a friend incapable of empathy or conscious reflection, and thus ideal for abuse.

The transition from human to robot butlers will therefore result in a vast erumpent outburst of wondrous new human freedom. With robot butlers to buttle us, we need not buttle ourselves. Of course, small concerns arise that this significant decrease in toil may be matched by correspondingly large increase in mass unemployment and misery. But as sensible free-market economists have pointed out, those who are thrust out of work by the introduction of the robot

butler can find new employment manufacturing and selling robot butlers. Seeing no flaw in this logic, we gesture rudely at those with humanitarian concerns.

SYSTEMATIZED TIME MEASUREMENTS AS INDUSTRIAL CONSPIRACY

But the age of industrial progress has transformed us in more ways
than one. It has not just given us digital handcarts and electrified
chairs, and allowed us to look at pictures of Maui instead of ever
having to visit it. It has also modified our very psychology, mashing us
into its own image. The combination of technology and industry has
constructed the very way we see the world, from why we use a comb
(to impress the boss) to why we modify our syllabi once a week after
the semester has already begun (so that our personal computers will
not feel neglected). We live in no place more often than Time itself,
however, and Time is one of the foremost products of a techno-in-
dustrial point of view.

By now, even the most hardened and crusty of our time-scientists
have conceded that our hours and minutes constitute the abritrariest
of all arbitrary measurements. Time, like all constructed systems of
quantification, must be designed to serve the ends of living. Our current time-system is a product of the cold rationality of the Industrial
Age. It is the time of the shop foreman and the office manager. It is
a time that measures human worth by "efficiency" and "productivity"
rather than "shimmering dynamic Life-energy" or "brilliantine radi-
ating zest." It is difficult to dispute that contemporary time measurements, divided into systematized units, are little more than an industrial conspiracy to suppress the jagged, multifarious, and local in favor
of the vast, impersonal, and centralized.

Thus, it is plainly time (excuse the amusing pun) for a New System
of Temporal Quantification. Our proposed adjustment to the system
of Time-Measurement is as follows: From now on, the day shall be
broken down not into the Hours and Minutes of the factory-owner,
but the heartbeats of every working man and woman. A glance at the
diagram on the ensuing page will suffice to demonstrate the concept
in its rough outline.

Observe as follows:

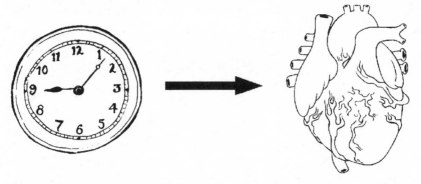

In this way, Life may be governed by its own rhythms rather than the staccato bursts of the hideous mechanical-contraption, which even at its most well-suited moments is one level removed from the essential controlling ticks and beats of mankind's flowing activity. The heart-beat measurement-cycle speeds up and slows down according to the needs of its user, meaning that a synthesis of the analysis of life and the components of life would finally transmogrify itself from Infeasible Vision into Unavoidable Actuality.

MECHANIZATION OF THE DANCE

It remains true that though it pitilessly destroys the hearts of all human beings, technology has given us tiner and tinier telephones. Perhaps indeed, however, our technology is a mixed curse with drawbacks equal to its uplifts. We always thought we would have robot-butlers, but little did we know that quite the opposite, *we ourselves would be the butler's robots.*

Technology is both hoorah and boo; i.e. we all love having our creases perfected by the local electric trouser press, but nobody wants to get his tongue caught in a wheat thresher. The automated hot-dog slicer is the house-husband's best friend until it carries away one's nephew's penis.

Thus, the downsides of technology must be considered. In this case, a particularly illuminating textual fresco can be assembled through a focus on culture: how have electronics changed the way we manifest ourselves in verse and motion? It is indisputable that every age's artistic output is solely determined by the state of its technology, and so how does this apply *here?*

Consider dancing. (Thought *do not* dance.) They say we live in a "New Age" (pronounced "newage" like "sewage"), one shaped by the precepts of free-love and loose association. But if this is somehow so, the question inserts itself: *Where are our rapturous dances?*

Recollect the dances of our pasts. The tango, the rabamba, the bamboula, the samba: though all were folly, each built up over the ages like cultural tartar in the corners of the collective body-mouth, until it became so thickly ingrained in our being-gums that you could not abort it with a planet-sized forcep.

One might then suppose that in the post-industrial age, with its heaping portion of free-form *je ne sais rien*, the evolution of the art of dance would by now have us tapping out our soul in rhythm-clogs up and down the deck of the H.M.S. Pinafore. It has not. *The Pinafore was our Titanic.* Passion and looseness of expression, plus the occasional forbidden glint in the sailor's eye; all are hallmarks of the art of dance and yet society has forsaken them, only to shamelessly abscond with the technological doodlebug. Does it not seem as if these days, when one wishes to take a girl to the dance, one first has to ask whether she even enjoys dancing to begin with? This was not always the way.

★   ★   ★   ★

It would take a jejune human being indeed to have *no* fear of external cyborg takeover. The countless leathery tomes and case studies[58] written on the subject are a testament to the fear's human ominpresence. Yet while all of these theories posit the future as a realm in which old Herb "H.G." Wells will have his last little smug giggle, even the most nuanced notions neglect[59] to sufficiently wildly speculate that such a takeover may have already begun, finding its nascency in society's sordid love affair with gizmology and the gizmological.

In each of our times, we have all of us seen a man or woman in a semi-professional halfbreasted suit walking down a city street in the daytime, mumbling to themselves with eyes downcast. If we could attribute the condition to the ordinary internal insanities it would give

---

[58]   See, e.g., *Political and Ethical Choices in the Design and Introduction of Intelligent Artifacts* (Whitbey & Oliver, Columbia University Press, 2006) and *The Robot Takeover!* (Universal Pictures, 1959).

[59]   The 3 N's of Legal Discourse

off no noteworthiness. But instead, said woman/man is inevitably engrossed in his/her digital telephone; *she is a mechanized man.* There is nothing left of her save that which is granted her by the omnipotent technocrats who design and regulate her Zune.

Our contention is as follows: the society in which all people have become cyborgs is undesirable, *whether or not that transition was nominally voluntary.* We used to complacently surmise that such a truth was self-evident, but have since learned that many not only fail to concur with it, but actively dispute it. The debatability of mass cyborgism strikes us as illustrative of the distance that the social locomotive has traveled away from the tracks and into the precipice.

But having stepped out for a cigar, we return to the dance. If mechanization can *im*prove certain spheres, such as the mass manufacture of cola-flavoured gummy candies, it can nevertheless *dis*prove others, such as dancing. If we believe that the sixth or seventh most important function of society is to create art, then we must have believed that the primitive gyrations of the human pelvis to the rhythmic beats of lost times were more liberated and spontaneous and thereby full of life than are our current digitized variations, whose "Dance Dance Revolutions" are the furthest thing possible from revolutionary.

THE TWIST

[Originally published as liner notes to an academic retrospective compact disc, examining the legacy of the 1950's "Twist" craze in light of post-Cold War cultural shifts. Though the text inevitably suffers in the absence of the accompanying soundtrack, we nevertheless feel as if failing to include it here would disserve both our readers and our word count.]

> *It is not true that doing the twist prevents aging, but it is true that not doing the twist prevents youth. Yes, daily America's view of the dance has been temporarily poisoned by a sailor's slew of cash-in LPs ranging from Bo Diddley's 1962 "Bo Diddley is a Twister" to Johnny Cash's same-year best-forgotten platter "Shot a Man In Reno Just to Watch Him Twist: 22 Dance Party Favorites from the Twister in Black."*

But lettuce recall an incident several years prior, when both Robinson and Nimni found themselves at that storied street-corner (Rampart and Canal). No, Charles Checker himself did not show, but what did arrive was revelation: a realization that without a good twist, we weren't going to get anywhere far. And so, as we began to twist, so did those elsewhere on the pavement, until by midday half the Crescent City was twisting with merriment and poise.

A symptom of too much masculinity? Perhaps. But we maintain that our twists, and theirs, were a community good, and to the skeptics we pose this question: is the presence of dancing in the streets not the foremost measure of a healthy city? GNP, GNH, DDT: all of these quantifications fail gloriously at capturing what it is we mean when we says "The society that twists together, stays together." Go out and find yourself a more accurate map of the Central Business District, we dare you. Number of twists per capita will continue to bubble to the top like a tableside Cousteau.[60]

MUSIC MORE GENERALLY

We have been accused of disliking music, because we referred to it once in print as "dessicated Dionysian ear-ooze," and issued unflattering remarks about the influence of Mick Jagger's lips on the scholarly vigor of today's youth. But nothing could be located more nearly next door to the truth. Though we have grimaced through our share of intolerably smegmatic concerts and concertos (a 1996 Austin, TX concert by *Law & Order* tribute band The Dick Wolves being an especially abject lowlight), sometimes we do not mind putting a record on and ignoring it.

But as they say, there's life, there's art, and there's *this*, and we confess to having toe-tapped to our share of harmonica-clad bandstand cornhuskers and beflanneled bluejeaned washboard-ticklers. "Life itself is a bluegrass!" one of us was heard to exclaim as we left the rec hall elated one midwinter's night in '82. "No, it isn't," we later realized, but that foul dictatress Nostalgia nevertheless keeps us from unleashing

---

60   Spatial limitations prohibit further explication of what we mean by this.

our full hatred on any musician wearing a tuxedo or strawhat.

Yet this does not mean that we are tolerant. We have very specific suggestions (orders) for how music not only *could* be made, but *must* be made.

First, all existing musics must be collapsed into one. "Genre" is exceeded in the arbitrariness of its boundaries only by the beard.[61] The glorious union of funk and country must be undertaken in earnest. A carnal relationship between hip hop and pipers' jigs must be forced at gunpoint. *No music should be produced that does not in some way incorporate or refer to all genres.*

But the precept most deeply lodged in the fundament is that music *must be more generously integrated into everyday life*, so that it is not the exception but the cornerstone. It should not be possible to spend an evening in one's bedroom without hearing a brass band pass by. Clerks selling chewing gum and cigarettes should always offer a complimentary serenade. Rail conductors will call to one another in song, belting old hymns about ticket-collection as they go about collecting tickets.[62]

Nobody would dare leave the house with anything less substantial than a zither in his satchel. The scourge of Muzak would be driven from the land; by law elevators would be required to offer live bands. The streetcorner doo wop group would become so commonplace as to be an annoyance rather than a curiosity.

Things do not end here; our intravenous injection of the musical into the mundane continues until bursting point. Engineering must become a music; architecture doubly so. Ladies who wear large hats should take care only ever to wear models that emit melodies when they are rotated on the head. A sideburn must ding when it is tweaked. Lovers should always give one another handmade music boxes as signs of affection, instead of greeting cards or kisses.

It becomes amusingly laughable, then, to accuse us of being unmusical. Certainly, our a capella rendition of "Shenandoah" at a faculty retreat's open mic night was met with both a mass walkout

---

61   See Appendix C.

62   Our own contribution to this badly neglected subgenre, the moving spiritual "Jesus is My Ticket to Ride," was lamentably caught up in twin frivolous copyright suits from both the Christian Church and the Lennon-McCartney estate. It was this incident that most passionately crystallized our staunch belief in the abolition of intellectual property protections.

and an official departmental reprimand. But that we cannot ourselves produce melodious birdsong does not mean we cannot insist upon it from others.

In many ways, however, this chapter is impossible. One cannot dance without a revolution, and one cannot violently revolt without a detailed historical theory of political change. Before restarting the music, let us instigate the revolution.

MODERN REVOLUTIONS

The generally received supposition is that revolutions are contrary to contemporary possibility. But the supposition is precisely that; a thing *supposed*. But suppose we supposed differently? Suppose we supposed that, in fact, revolutions are far from having taken their final inebriated bow and plummeted off the stage. That they are instead mere infants making fumbling mistakes, babes who must be given a chance to learn and grow after committing their first foolish genocide. Perhaps if we think carefully about what revolutions are to begin with, we may decide they can indeed be harmonized with an era of glass architecture and digital watches.

It goes without saying that revolution carries risk. After all, to cut off the head of one's leader is a radical act in itself. But to do so without quite knowing what one is going to do afterward, *this* extreme is the revolutionary's divine charge. The unwise consequences of such senseless beheadings need not be listed; one may end up eliminating the only individual with a full working knowledge of regional agricultural policy, or even accidentally murdering the wrong fellow altogether and having to apologize to an irate sibling.

We have always been cautious revolutionaries, then, if not exactly tenured ones. We recognize that the man who goes about calling for various deaths may soon be on the list of various deaths called for; in attempting to stab the ant between one's feet, one may end up slicing off one's toe. Everyone wants a grand Revolution, but nobody wants to end up like America did.

The problem, however, is that in a total rejection of revolution, the pendulum has swung firmly into the other testicle. Avoiding swimming in rapids for fear of approaching a Robespierran waterfall, we have decided instead to bathe in the stagnant water of reactionaryism.

But bogs make poor bathfellows; one need not choose between getting dashed upon the rocks and getting leeches in the navel. To realize that political radicals are a terrifying bunch does not entail becoming a pallid and lifeless political moderate. A whole ocean of arousal swims between the ever stiff and the permanently flaccid.

We became far more convinced of our own position on a recent trip to Cuba. After discussing some conference notes over *batido de pinas* at the Hotel Nacional, we happened to encounter a dignified traveler in the expatriates' lounge. This *hermano* was a bespectacled revolutionary of the *escuela Viejo*, a true *dignitario*. He related tales of anti-Batista struggles from long past; in his telling, Batista did not come across an especially endearing man.

When he concluded his good-natured storytelling, we were taken aback by his and his people's courage. Yet we found ourselves having to say to him: "This revolution is all well and good, but how would we do such a thing in a stratified and federal society like America?" Our revolutionary replied:

"Hay los montañas?"

We answered that we did.

"Venez a eso; pagez una contigenta armas." [63]

Skepticism flooded forth. Such dreams were outlandish! The practicals were impossible! Yet during the reflections upon the man's remarks which ensued over the next couple of days (as we recovered in our *camas* from a particularly debilitating *noche de mambo*), we began to reconsider ourselves.

Ultimately we concluded that revolution could indeed be made. But it would not be conducted, as our *conocido* had suggested, with remote-controlled pistols and cantilevered machetes! Instead, it would be a revolution of subtlety, grace, and restraint. A revolution of the spiritual and scholastic instead of the material and real. It would be unlike all prior precedents. It would indeed be televised. It would be sympathetic even to its victims. It would understand the crucial inter-connotantional differences between "thing" and "thang," and it would give each human brain an equal share in the National Consciousness. It would be a countrywide Stock Exchange of citizen potential and productive mindwaves!

---

63   Trans: "Do you have mountains?" [...] "Go to them; form an armed contingent."

Our bearded *companero* had thus made a small error of reasoning in his substitute of Guns for Butter and Butter for Brains. Mountains will be gone to, but they will be the majestic mountains of mental meanderance, rather than the cold, stony peaks of the earthly ranges.

Revolution, then, must be reconsidered. It is not enough simply to behead; *one must also know what one is doing.*

CONTEMPLATE THE TWEEZERS

But perhaps we would spelunk further by using metaphor rather than fact. Sometimes a physical object better represents an idea than the idea itself. For these, we will deploy the medium of the bullet, which has proved an effective means of addressing the majority of global crises:

♦ *The tweezers are the foremost weapon in the revolutionary's toolkit.* Unparalleled in their elegance and refinement, the tweezers were one of the earliest devices to spring from the mind of mankind. The butter knife, the ox-churn, the claw-hammer, and the diving bell all appear crude and obvious contraptions when placed along-side the sublime usefulness of the well-tuned tweezer.

♦ *Liberation itself is built from Tweezing,* from the disman-tlement and dissection of everyday reality. Bit by bit, it takes apart the diseases of the human situation, and indeed, the situation itself.

♦ *The Problem of Politics is that it necessarily deals in Aggre-gates and Abstraction, ignoring the Individual and the De-tail.* The tweezers are the opposite of politics. They cannot help but remember the Detail, for they are constructed *in order to* deal with it. The splinters and unsightly hairs of our nation must be plucked, not legislated against.

The mechanics of revolution have thus been carefully elucidated. But what of its justifications? When is one permitted to rise up and throt-tle the landlord, and when is this improper? We must lay the moral foundations of rebellion if we intend to build an aviary on them.

ON THE SUPPOSED RIGHT
TO REBEL BECAUSE OF NOBLE CONCERNS

Time and again our tiny yellow planet has been shaken by upheaval and rebellion. Logically, questions of necessity and revolt ought to occupy the central position in our cultural lexicon, and be the primary subject of our many panel discussions. Yet they fail to, and we are left aflounder in the sea of perhaps regarding whether or not a revolution is justified at any particular intersection. To this chapter, then, the job of organizing a comprehensive normative ethics is left.

However, *general theories of rebellion* are too intricate and require too basic a knowledge of history for us to discuss here in their grand context. Were we to attempt to illustrate moral principles using historical examples, it is no doubt we would be deluged with "corrections" of our "inaccuracies" by lunatic members of the scholarly fringe. Instead, to learn safely about rebellion we must examine it on a smaller scale, using casual fictionalized mini-scenarios, knowing that the truths we discover can easily be woven into a more general theoretical tapestry, and by their nature cannot be subjected to factual quibbling.

Imagine there is a classroom, in a school, in a semi-sleepy suburb. In this classroom is housed the school's music class, taught by the teacher: Ms. Bell. Bell has one rule that must be adhered to strictly: students are only allowed to go to the bathroom when she allows it, and only permitted to spend a maximum of thirty seconds within. This rule, and its corresponding fear of punishment, has had the general effect of significantly reducing the amount of children that go to the water closet during her class. However, on occasion a child simply *must use the lavatory*, and it is his/her case we will examine further.

Let us look at the experience of child K. K has to urinate. However, for fear of exciting the wrath of his teacher, he has conditioned himself to hold his urine. Today he cannot. Therefore, he urinates on the carpet. The other students do not laugh, as they all experience the same trials, but nonetheless K is wet and uncomfortable. He then picks his nose and defiantly wipes the contents on the carpet as well.

This additional action may seem gratuitous, but it should be looked at more thoroughly, through the lens of political theory. The fear inculcated by Bell led to *both the unhealthy conditioning of K and to the urinary incident*. This is fairly apparent. Yet it should be argued that

K was *obligated* to commit his further nasal action as a demonstration. The urine itself *cannot* be considered an act of rebellion because it was unconscious and merely a result of tyranny. The nasal action, however, is justifiable because *the mere urination does not exhibit discontent with an existing policy*. In addition, few can rally behind a urinary flag, but by backing outright rebellion we can hoist the banner of defiance high.

A Religious Temperament

Being religious without God is thought of as akin to being wet without soap. But those who say this misunderstand the nature of soap. Soap is not actually designed to *heal* us, but to *make us clean*. Yet is it proper to call God "clean"? Can mortal conceptions of filth even be applied to divine epidermis?

In some ways, it *is* theoretically possible to shower using water alone. And so may it be possible to be religious without subscribing to the bi-weekly podcasts of any Gods. For to be religious is fundamentally to believe that there are transcendent what-have-yous, and transcendent what-have-yous do not find themselves necessitating the life-giving touch of Godly lips.

Each of us has occasionally been spurned by a prelate or devoured by a vicar. But is that a reason to sour on religious candies? Jesus, it must be remembered, was not an intellectual. His ambassadors are not himself; your average dockside dung-unloader more livingly embodies his Spirit than all the truckloads of archdeacons on all of America's highways.

There can be redemptive value in religious tradition. Christmas, for example, is a lot of fun. The question for those who do not believe any of the theology but enjoy the part where we sing and receive presents: can there be an intellectualized reconstituting of the Religious Event? To put it another way, what happens when Christmas goes to college?[64] The ritual of celebration, consistent across ripened human cultures from the Samoan to the Londoner, serves a purpose,

---

64   "Christmas Goes to College" is used here in neither of its traditional meanings as the title of a holiday-themed erotic postcard or the pseudonym for a certain type of vile fraternity prank. Instead, if it is not completely clear, we are on the subject of the rôle played by the Holiday in catalyzing human progress. When we say it is time for Christmas to finally grow up and go to College, let nobody smear obscene misinterpretations upon our gist.

maybe. It would be a shame to rid ourselves of mystification, but at the same time having to resign ourselves to never again hunting for an easter egg, transubstantiating a madeira, or eating a cookie shaped like a crucifix. A religious temperament, then, should be kept, even as our religions themselves are heaved into the offal barrel.

Never Letting Atheism
Come In The Way of a Good Party

But it would be senseless to discuss religion without diagnosing its opposite. Whenever there is a belief to be believed, there is a belief equally well to be disregarded. Thus, the Atheist.

The Atheist has a problem, namely that he is a pompous drag at barbecues. Every time we have experienced the Atheist in our social lives, whether at a faculty cocktail evening or a faculty wine retreat, he has done little more than carp about the primacy of Reason and the folly of unproven theisms. No number of gentle asides informing him that he is behaving noisily and dung-headedly have managed to lower his volume.

The main question surrounding disbelief, then, is not *why does God not exist,* but rather *why does the Atheist insist vigorously not only on being right, but on doing it loudly?* Of course, we are under no illusions that answering this query will resolve all global inter-sectarian tussle. But we do believe it will result in a vastly more satisfying Friday afternoon *hors d'oeuvre* hour in the Sociology Department lounge.[65]

With that, we believe the issue can be put aside. Kindly do not get us started on what the Vegan does to a party.

Honesty

But let us turn from the specifics of atheistic baseness to a broader meditation on qualities of character and integrity. The question of when one should be honest and when one should plagiarize is an interesting one. Obviously, it is impossible to be honest almost all of the time, but is it nevertheless possible to be honest with reasonable regularity? And if it were, would we even desire it to be so?

---

65   This Atheism section of the book should in no way be considered to be have been written solely in order to disparage a particular departmental colleague.

Consider this newspaper:

> *SPRINGFIELD -- The last name of Green Party guberna-*
> *torial candidate Rich Whitney is misspelled as "Whitey" on*
> *electronic-voting machines in nearly two dozen wards -- about*
> *half in predominantly African-American areas -- and elec-*
> *tion officials said Wednesday the problem cannot be corrected*
> *by Election Day.*
>
> - <u>Chicago Sun-Times</u>, October 14th, 2010

At the time of its collapse, the candidacy of "Rich Whitey" may not have had the level of popular support it deserved, but we feel that the story requires retelling, if only to provide a crucial honesty lesson for generations of sticky, credulous children to come.

We know that, in a sense, all candidates are Rich Whitey. But only Whitey himself was willing to come out and acknowledge the fact, brave fellow. The incident scarred his loved ones, but did he shirk? He did most assuredly not.

This kind of candidatory straightforwardness is necessary for both those that wish for the destruction of the current political system (such as us) and those that wish for its preservation (such as them). If candidates are honest, those that revel in the republican process will be able to place more trust in it than ever before, but those that despise the republican process will see its full horrors exposed for all to see and no longer concealed in a web of squelchy rhetorical emptiness.

SAYING WHAT YOU MEAN VS. MEANING WHAT YOU SAY

But perhaps we cannot be honest without understanding what we mean. And we cannot understand what words mean without a robust theory of the construction of the Authentic in modern discourse. After all, the crisis of inauthenticity in our schools continues unabated. Woe betide the noble professor who steps to his lectern and mouths the Truth That Cannot Be Spoken: "You pupils are nothing but automatons! The words you spout have no meaning behind them!" We know full well from copious experience what becomes of such a daring truth-teller; this very work is the partial result of one such forced sabbatical.

Yet it is correct, isn't it, to say that young people of the present age haven't a clue what they are saying and definitely do not mean it? Go out and listen to one. Imprison it in a box and attempt to comprehend its yelps. It's an absolute certainty that nothing will come of such an enterprise - for this is a new age of antipathy. What does it mean to mean what we mean?

There are two ways to tell the truth. The first way is to say something that is true. The second way is to say every single thing that is not true, so that the truth emerges in the spaces between. But this gives the impression that the truth and its opposite are somehow the same; that is not the case. For by telling the truth, we produce only mundane conclusions, statements whose value is close to null. We produce phrases like "I am not myself today," or "If you can't get at least four students to enroll in this Utopian Torts class, there's really no way we can let you offer it." But by telling every possible lie on the route to the truth, we get something far more palpable.

Just because lying is a form of telling the truth, however, does not mean it does not remain vital to be extremely serious about it. We have always cautioned our contemporaries against the risk of failing to adequately take things seriously.[66]

But *semantics do not play politics*. All politics might be semantic, but there is no use in reversing definitions, for a man cannot exist if he is turned inside-out. *Nothing comes from a mumble but a breeze.*[67]

---

66 A brief but necessary digression: We recently wrote a lengthy manifesto on the tendency of academics to insist they are taking their subjects seriously. Though the full text of the masterwork itself was lost in a tragic accident (confusion of the CTRL-cut function with the CTRL-paste function), we nevertheless managed to preserve a portion of the uncompleted draft introduction to the article (with our own annotations and notes to self), entitled "Taking 'Taking Seriously' Seriously." We reprint it here for its future historical value:

"*[...] It is understandable that so many academics would insist that they are taking seriously the hitherto-frivolous. Nobody wants to be accused of being lighthearted or having a sense of humour. Yet in taking all of these subjects seriously, the aforescholars have forgotten the most crucial truth: to be a scholar is inherently to take seriously. We have done our job in this article; we have written it. Yet writing is not interpreting, lest it be forgotten. [INSERT TRANSITION HERE, PLUS SOME CONTEXT] The uneducated reader may question our motives with the perennial dunce's refrain of 'But is that legal?'. We for our part have established legal records to sooth these soothsayers, but records are of course B-Side the point. [DOES THIS PUN WORK? SEND EMAIL ASKING THE CLASS] Everyone wants to be taken seriously, and everyone is. Such is the problem. [USE SOME HUMOROUS EXAMPLES] The circumnavigation of one's own colon is never an easy task, but it will be worth it if at last we can stop taking Subject X seriously, and at last begin taking taking seriously itself seriously.*"

67 Aphorism #14.

And so it is we come to...

TERMINOLOGICAL DISTINCTIONS

We have long been prized for our addiction to distinction. A recent book on the subject put out by the editors of *Porous First* named our syntactical treatise *Seven Horrors of Linguistic Eclipse*[68] "one of the most attentive" contemporary inquiries into the political implications of careless word usage.

¡Do not mistake our position! We are not the tweedy grammatical pedants some would mistake us for in the dark. We reject nearly all contemporary debate regarding word usage and usages. But that said, we feel that semantic caviling has an important and fresh-baked roll to play in sorting out the human muddle.

To use a rather childlike example from the annals of water-fowl, to "quack" is not the same as to "engage in quackery." But to "shag mercilessly" and "to engage in a merciless shagging" *are* synonymous. Why the distinction? Praise be to Burgess, who first felled this partic-ular linguistic oyster with his Big Blue Theory: *we draw the distinctions we need, whenever we happen to have need of them.*

These days nearly everything is misunderstood. We well remem-ber the time an academic colleague asked us for our opinion of his new foreskin. We told him we thought he could pull it off. We meant, of course, that on him it looked fashionable, but our compliment was tragically misunderstood. *Ambiguity is the mother of circumcision.*[69]

But if that was a case of too few distinctions, there are equally often too many. We turn now, then, to the related question of how to interpret the difference.

---

[68] Originally "Linguistic Equipoise," changed at the recommendation of an editor suffer-ing the mental illness of grammatical prescriptivism who believed delusionally that a "ui" formulation could properly appear only in non-consecutive words. This sham rule appears nowhere in either the *Kellogg's Dictionary of American Usage* or the *Default Usage Example Text.* We thus take the opportunity of this footnote to point out just how correct we were, and what an injustice was wrought when we were prohibited from submitting to future PF issues due to the minor physical altercation that resulted when the afore-mentioned editor was vigorously confronted over this very grammatical point. It should be noted that this decision was not altered even upon our full reimbursement of the instigator's grossly inflated medical expenses.

[69] Aphorism #15.

Hermeneutics

Not all words are equally long, thus not all words are equally useful. Small words are capable only of capturing small thoughts; each additional syllable adds a new layer of meaning and scholarly depth.

By this reasoning, perhaps no word has done a greater service to the academy than *hermeneutics*, which both clarifies and describes.[70] Not only is the *sound* of the word pleasing (and playful, suggesting incorrectly that the subject under discussion will somehow relate to newts), but it is among those rare terms that have actually improved every single piece of scholarly writing in which it has been included by an author.

There will be no need here, of course, to define the word for readers, or to provide examples of how and why it is useful. In assuming that our audience is worldly and learned enough to know these things, we trust that they will appreciate our lack of condescension. To systematically make the case for why hermeneutics are meaningful or important would insult all of those who have shelled out valuable pocket-change for the privilege of showing others that they possess a copy of this book.

But all these mentions of newts remind us of certain reptilian social theories that must urgently be posited. Let us proceed now away from language and towards the snake.

The Necktie as Serpent

Have you ever witnessed men wearing ties that have the image of a snake plastered upon them? This is not what will be discussed here. What will be discussed is the way in which the necktie, in all its *machismo* and professionalism, has come to represent something far more sinister than a mere button-disguiser or genital-enhancer.

The necktie is the very *serpent of our age*, the damnable temptress that lures us from that which we know to be Pure and Straightforward

---

70 We temporarily put aside the issue of whether it can ever be appropriate to "service" the academy in public; we have become somewhat more cautious on the topic since our calculated use of this suggestive pun in a syllabus led to four baseless parental complaints and a stern email from a department chair. Hypersensitive halfwits like the aforementioned parties are discouraged from reading *Blueprints for a Sparkling Tomorrow*, though they are nevertheless strongly encouraged to purchase copies for ritual burnings.

to that which is Menacing and Insidious (to put this transition in the framework of a helpful metaphor, the Tie takes us from the Ambassador's Ball to the Sinful Hot-Jazz Nightclub, where scandalously-clad anacondas bare their leglets for money).

Yet in spite of the necktie's subversive and seductive power, it has somehow managed to become co-opted as a symbol of all that is painfully grey and rational in the world of the Businessman and the Bureaucrat. How can a snake so potent become a garment so dull? The query is rhetorical, but its curiosity rings far too true. The answer is unknown to us, but we do know the solution. We must take back the necktie, and restore to it its symbolic power as a Rascal and a Deviant. The necktie must not allow itself to become the next Cufflink or Neckerchief. It must reassert it libido through vibrant color and sinister activities. Rakes and libertines must wear them as they practice acts of gross sexual indecency before gatherings of shocked and affronted aristocrats. Motorcycle gangs must sport them as they beat one another with pool cues or bludgeon fans to death at rock concerts. And princes must wear them as they ride gallantly into battle. By restoring Cool to the necktie, we restore the Spirit of the Snake to our social order.

THE SERPENT AS NECKTIE

We believe that the serpent is criminally underutilized as a necktie-substitute. Consider the ancient piece of Turkish Wisdom:

> *A man whose tie carries potentially lethal fangs is a man who commands respect among his peers.*

Yet when it comes to snaketies, how many among us have carried our principles into practice? Fear of bites and blemishes has displaced our innate bodily need for masculine fashion. Even Dr. Theodore Nugent is willing to entertain the wrath of the both the Secret and the Selective Service, yet has thus far proven unwilling to replace his Pierre Cardin with a writhing python. The *feather boa* has achieved a great deal of popularity on the nightclub circuit, yet the *boa constrictor* languishes in our swamps and bayous, friendless and neglected.

Given the crisis, we appeal to you, the humble proprietor of a

Tie Rack outlet store or franchise, to ask your supplier about snakes. As you were informed upon acquisition of your business, they are traditionally excluded from catalogs and base stock due to their general unpopularity and handling difficulty. Yet sometimes we must prioritize the *interest of the nation* over short-term gain. The Gentlemen of our Earth are in dire need of hissing, spitting serpents to complement their business wardrobes, and who are we to deny them this freedom?

Owl Pellets and Democracy

Reptiles are not the only class of animals with important political metaphors to offer, however. Consider the pellets produced by the latter-day owl. Each contains fur, bone, and additional extraneous mouse-bits, and is expelled from the animal's proventriculus in order to save the contents from a most uncomfortable journey through the owl's more fragile digestive pipes and tubes. The owl pellet is the pre-filter for all undesirable substances.

Our democracy has no such pellet. In Justice Holmes's conception of the "marketplace of ideas," even the most tiresome or despicable proposal can be placed into the collective deliberative square and put up for consideration. The pragmatist allows nonsense, insolence, and Bolshevism to proliferate, on the hopeful theory that justice will miraculously emerge from this swirling torrent of incoherence. We refuse to impose value from above because we believe it will emerge spontaneously from below.

All of this is well and good. It represents an adorably trusting view of mankind's potential for self-governance. But it does not confront or contemplate the Essential Question: What happens when one of the system's inputs is corrosive of the system itself? What happens when the marketplace not only fails to produce Justice, but collapses in on itself, crushing all of those poor souls who had the misfortune to be standing beneath crucial support beams or pointy chandeliers at the time of the implosion?

Consider television and its effect on justice. In recent years, legal professionals have begun to whine that the procedural crime drama has had a measurable effect on the deliberation of juries. Jurors are absorbing and regurgitating the investigative tactics they see on Perry Mason and now "think they have a thorough understanding of

science they have seen presented on television, when they do not."
Criminal defendants' right to a fair trial is therefore being eroded
thanks to Perry's right to free speech.

So what do we do when our very democratic system is threatened
by one of its own limbs? If one of our hands attempted to strangle
us, or one of our feet tried to administer a kick to our gonads, would
we remove it from our bodies or let it pursue its happiness? The answer may lie in Owl Pellets.

The owl has a democratic system of digestion. It allows many
things to pass through its organs, retaining what is useful and expel-
ling what is not. But the owl recognizes that certain inputs are too
harmful for the process at large to be trusted with. Defecation alone
is not a suitable method of expulsion. Instead, there must be a preliminary process. The owl does not refuse to *consider* the bones of the
mouse, but they are the first items to be spewed back into the world
through the inner beak.

In the same way, a marketplace of ideas must have a filtration
mechanism. It cannot fail to initially consider awful television pro-
grams or Fascist Ideology, but it must dispense with them rather
quickly, lest irreparable damage be done to the minds of younglings.
We must create a pellet of our most corrupting and deadly thoughts,
and toss it into the sea. *Only the pellet can guarantee liberty.*[71]

---

71 Aphorism #16.

C. Political Arrangements

THE CONSERVATIVE DISPOSITION

We wear no sandals, sport no beards, yet we have consistently been labeled members of the "extremist ultra-left set" by those for whom idiocy is an aphrodisiac.

This criticism ignores the profound conservatism of some of our most potable ideas. It is true that we may have not undergone the now-traditional ritual Baptism in a pool of Barry Goldwater's tears, but we did quote him in a magazine once.[72]

What does it mean to show conservatism in the face of rationality? Michael Oakeshott sprouted the following proposal-plant:

> T[he right] centres upon a propensity to use and to enjoy what is available rather than to wish for or to look for something else; to delight in what is present rather than what was or what may be...Since life is a dream (yet is not), we argue (with plausible but erroneous logic) that politics must be an encounter of dreams, in which we hope to impose our own. Some unfortunates, like Pitt (laughably called "the Younger"), are **born old,** and are eligible to engage in politics almost in their cradles; others, perhaps more fortunate, belie the saying that one is young only once, they never grow up. But these are exceptions. For most there is what Conrad usually called the "shadow line" which, when we pass it, discloses a solid world of things, each with its fixed shape, each with its own point of balance, each with its price; a world of fact, **not poetic image.**

---

72 See "The Conches of a Conservative: Politics and the Art of Seashell Collecting" by Oren Nimni & Nathan Robinson, *Beaches Monthly*, April 2000.

Oakeshott, of course, appears to have misinterpreted us. But his is nevertheless the kind of conservative that we find ourselves, the kind more interested in conserving than in causing conservatism. Enjoy what we have instead of crumpling it into a ball to origami it anew. Do not make waves, for waves rock boats. Of course, there must be exceptions to this rule. Progress cannot be rejected each and every time she asks us out for a drink. One need not become intoxicated with her in order to realize that she can be jolly good company. A love affair with progress can be a society's ruin, but a cordial standing date with progress might be just what a society needs to perk it up from its lingering postwar funk.

Nobody who reads *Blueprints for a Sparkling Tomorrow*, even casually, will be under the impression that we do not believe there are tinkerings to be made in the underparts of the contemporary social apparatus. These are, after all, blueprints rather than artifacts.[73] But believing that there are things that ought to be done does not mean believing that all things that could possibly be done are thereby ripe for the doing. The format of this book, which contrasts those things which must be preserved (Part I, Orthodoxies) with those things that must be discarded (Part II, Incompossibilities), is a successful attempt to structurally elucidate the principle.

There are some useful medicinal fungi growing in the damp bog of conservative thought. One need not be as frightening as that storied New Mexico governor who campaigned as being "more right-wing than God" in order to sensibly poach a conservative precept here and there. There may not be much of salvageable worth in the beached husk of today's Republican Party, but the principle "try not to bayonet the old traditions until they truly deserve it" is a worthwhile one to keep in one's mental apothegm jar. We may pocket it, and discard the rest of the conservative intellectual tradition over our shoulder like a de-frosted cupcake. But if that is the sum total worth of conservatism, what then of socialism? Does it, too, have some lasting juicy offal clinging to its abandoned carcass? Perhaps a clue may be found on the beach.

---

73 Readers interested in the difference between a blueprint and an artifact are encouraged to contrast the present work with our co-edited collection of facsimile 19th century literary sleeve art, *Now That's A Frontispiece!: 32 Undiscovered Woodcut Gems in Full-Color Reproduction Plates* (University of Morristown Press [Special Collections], 2009)

THE SOCIALIST BEACH

If pennies were administered every time someone had proffered to us the bendy old saw that "beaches have no politics," the resulting penny-administration oversight bureaucracy would rapidly grow unmanageable. Of course beaches have politics; even the manta-ray has politics, and it is a manta-ray. (The manta-ray is a Platformist.)

Everything has a politics, if one only takes the time to peel back its jumper and put one's tongue in its navel. And so it is with everything, so is it equally with the beach. Each grain of sand is its own tiny vote in one direction or another.

Consider a beach that is free to all. Mom and Pop decide that this morning would be a lovely day to take Gramma and the kids to look at some oceans. So they pile umbrellas in the jitney, pack a cooler with lemonade and saltines, and tootle down Beach Road to meet the sand. Now, in your ordinary city, this is fine: the family goes to the beach, parks at the beach, sits on the sand, and thoroughly enjoys itself.

But consider the unspoken premise: socialism. For Mom and Pop are poor, but we have not needed to mention this. It has been *immaterial*, because the beach is free to all. Luther Pennybags IV makes precisely the same sandcastles as Billy Ray Workingclass. A public beach is nothing more than a long strip of fine white communism.

That said, there are unsocialist beaches in the same way as there are unsociable breaches. The people of Connecticut, whose universities are known primarily for failing to renew the teaching contracts of their most valuable instructors,[74] have been known to charge exorbitant admission prices for the very privilege of gazing briefly at one of their inferior beaches. One can see what happens when the distribution of sunshine and waves is left in the hands of hedge fund managers.

---

74 We refer here not to the case of a certain "radical anthropologist," whose forcible ejection from a certain unnamed Ivy League institution we met with a smug guffaw; this roly-poly Anarchy Dad should have realized that one cannot cloak oneself in ivy and subsequently eat it, too. Rather, the case under reference in the above clause is that of two assistant instructors who were roughly informed by a department chair at "Wesleyan" University that multi-year sabbaticals could not be granted to faculty of the lower ranks, no matter how significant a contribution to geodesic scholarship the resulting written output promised to make. That these two professors were informed that their defiant taking of such a multi-semester siesta was grounds for the cancellation of their Hegelian Sexualities seminar is one of the most overlooked incidents in the suppression of scholarly liberty in the recent history of the American academy.

Thus, the socialist beach must be where we stake society's vast multi-hued umbrella. Not to recognize the revolutionary import of this most American of institutions means leaving ourselves open to a regime in which we are charged a dime for every additional square foot of sand we wish to traverse, or in which predefined rectangular sunbathing plots are allocated by competitive bidding process, the proceeds of which go to Shoreline Resource Management, Inc., a company whose board of directors has little understanding of the simple pleasures of experiencing universal free parking or watching a destitute child laugh as it is bitten by its first sandbar crab.

EVOLUTION

But socialism is not just fashionable; in some sense it is veritably *fashion-derived*. After all, can it be it any coincidence that the current President of Bolivia both sports an iconic alpaca sweater and oversees the only quasi-socialist state worth living in? Ever since the inauguration of the Morales regime, we have wandered our campuses in puzzlement, and pondered the quandary at the pond in the quad. The Evo Morales revolution (evolution) has been an inspiring example of the Left in action, which should enable us to proudly wear socialism's fuzzy sweater once and for all. True evolution is not some incomprehensible theory about animals, but a statement on the way certain kinds of fashion choices might enable certain kinds of political reality.

Capitalism

Socialism

Lest the situation necessitate further explifying, witness the ensuing quotation:

> *But when Mr Morales wore the sweater for his meeting with the*
> *Spanish king, Juan Carlos II, the muttering soon began. "Is*
> *there no one who might lend Mr Morales a dark suit?" asked*
> *a writer from Spain's conservative ABC newspaper. Others*
> *sprang to the defence of the new leader. The sweater, declared*
> *Manuel Rivas in El País, is a "knitted declaration against*
> *invisibility". Mr Rivas said that while generations of Latin*
> *American leaders have worn the policies and the clothes of the*
> *IMF and the World Bank, Mr Morales was showing that he*
> *was a man of the people.* - The Guardian, 20 Jan. 2006.

Can there be any remaining doubt that it is fashion, rather than politics, that comprises the political? The written and armed declaration
have historically been given priority over the knitted one, but as the
above diagram illustrates, this could well be a mistake.

IN PRAISE OF DUBIOUS ECONOMICS

Because our training is in social theory, it has always been difficult
for us to understand basic economic precepts. We have nevertheless
written several books on the subject, most notably *The Hand in Which
the Penis is Not: Markets and Economies of Scale in Postmodern Capitalism*, in
which we suggest that the invisible hand is nothing of the kind.

However, since we subscribe to Economic Reasoning, we know
that each economic activity that takes place within the walls of our
grand nationwide city-state enlargifies its Gross Domestic Product.
This applies to acts of sickness as well as health, murderousness as
well as peacefulness. Knowing this as we do, we push boldly here for
a new economic policy. We (and many of the more *starry eyed* economists join us in this recommendation, though mainstream ones do
not) suggest "Economic Growth Through Micro-Violence."

We recommend the proliferation of fists forcibly inserted into
the stomachs of the unwitting,[75] and the adoption of such a posi-

---

75   This does, of course, contradict Aphorism #17: *"Caress the unsuspecting."* But desperate
     times and such.

tion as State Policy.[76] This would include the financial subsidization of brawling on a state level, and the encouragement of films that romanticize the victimization of bystanders. The motion picture industry must finally begin to depict violence as being exciting and free of consequence, instead of constantly producing blockbuster after blockbuster focusing solely on its long-term costs in human suffering and trauma.

The formula is simple. Other than the mechanical trades, such as Ferris-wheel repair, plus self-publishing authors, hospitals are the most expedient means of GDP inflation available to the modern civilized government. And the best and most economical way to get the populous committed to the new system system should be obvious: a series of alternating light and heavy thrusts to the abdomen. *Violence remains as persuasive today as during peacetime.*[77]

The question still percolates - why is this the *ideal* viable economic stimulus rather than some other that better makes use of the selfsame resource-material without involving wounds to the gut? Might there not be some undetected fallacy in a philosophy that takes the breaking of windows for an optimal allocation of one's time?

One answer could lie in the classic tale of the Worcester Mariner, who continually wondered why his fish were too short, until he realized his ruler was broken. The fault may not be in the perpetrators of violence themselves, but in our systems of quantification and incentivization. If we would abandon this foul and amoral measurement mechanism, and switch over to a more cordial success metric such as Gross Domestic Love, perhaps we would see levels of barbarism and imprisonment decrease, and levels of pleasantry and smoochiness skyrocket.

THAT WHICH CANNOT BE QUANTIFIED

Throughout this text, we have tried to spare our reader charts and graphs, which make unfair demands of the audience by *showing* rather than *telling*. Yet the graph is the universal medium of the time, used to document balances and imbalances alike. Very few objects or inter-

---

76   At least, more so than is already the case.

77   Aphorism #18.

actions have remained ungraphed since the invention of the pocket computer. That small quantity of items which has thus far managed to escape the ravages of the graph-typhoon will find itself given axes and scales just as soon as the statisticians return from their quarterly regrouping-siesta.

We take mild issue with this storm of graphings, however, due to its useful but incorrect supposition that all things may be quantified, and that the Ephemeral and Metaphysical are mere pseudo-religious poppycock. The Economist believes he can precisely measure the Inputs, Outputs, Supplies, Demands, and Equilibriums of *love itself*, which we feel is a position drastically out of step with the reality of his position within Mother Universe.

The successful businessman may purchase a vanity license plate that reads NMBR1, he may cauterize his wounds with capital, and he may ride the finest of well-bred mustangs, but until his methods of measurement are adjusted, he is as blind as the man with no eyes. No matter how many subprime mortgages he pushes on how many teary-eyed orphans, he will not be able to reach even a basic understanding of humankind's meta-economic operations, let alone Geodesic Nirvana. Until we rewire our measuring-sticks,[78] we are compassless, morally speaking. Justice could be tossed in the hamper and we would be too blind to hear her muffled pleas. To get her out requires a whole new set of quantitative extraction tools.

Draw a graph of Vague Disquiet! Chart the flow of True Understanding! You cannot, and must therefore abandon your methods in favor of a numberless yet infinitely more accurate Concave Introspection.

---

78   For programmatic suggestions as to the effecting of this, please see our booklet *The Socio-Rectal Thermometer: Lessons for Economists from the Sphinctral Sciences*, in which we point out the absurdity of a world in which one's rectal temperature can be taken daily, but in which it is considered nonsensical to speak of measuring a society's temperature similarly.

IV. GROWTHS

"Progress is the towel that rubs us dry. Each soft cotton flick of progress can penetrate the darkest, dampest corners of our mired and filthy selves, and polish us clean."

- Prof. Fry (attributed)

"I wanted to run after him, but remembered that it is ridiculous to run after one's wife's lover in one's socks; and I did not wish to be ridiculous."

- Leo Tolstoy

"While the aristocracy was in the ascendant, patient hirelings used to apply their knowledge of hydraulics to the working of fountains, as in Versailles, or they devised automatic chess-players, or they contrived elaborate clocks which struck the hour, jetted water, caused little birds to sing and wag their tails, and played selections from the operas."

- Mumford, <u>The Golden Day</u>, p. 41.

What will Tomorrow's Future contain? We have speculated in previous chapters on its implications for the Stereophonic Defibrillator and the Unmeltable Creamsicle. But we have not yet taken a textual photograph of tomorrow's town. Yet from the current literature in futurism and speculation, we believe we have some small, grey inkling of the characteristics of the next great urban landscapes.

It is all-too-common to assume that humankind has nothing to look forward to but the enormous cosmic pratfall of self-extinction. The number of doomsayers seems to multiply hourly; nobody dares consider what life ought to be like in five hundred years, when we are so doubtful that our species will live out the afternoon.

We reject death-obsession as unproductive. The time we spend contemplating the various ways in which we could destroy ourselves is time we could spend dreaming of elaborate new contraptions and philosophies. What became, we sometimes ask ourselves, of the idea of having something to look forward to?

We count ourselves among the forward-lookers, and so in this section we think about what ought to be, if we can manage somehow to pull it off. The acknowledgment that doom is almost certainly inevitable does not mean we must think of nothing else. "Almost certainly" is not "certainly," and if we miraculously manage to perpetuate ourselves for some considerable amount of time, it is those who came up with *Blueprints* to suit the contingency who will surely be most in demand.

A. Structure and Infrastructure

BULDING A MOUNTAIN

"The youth of Kansas [should] build a mountain,
so they can have manly work and enjoy skiing."
- Paul Goodman,
Utopian Essays & Practical Proposals,
Random House (1963)

If this human race is to make progress, it must immediately and
enthusiastically adopt a new view of what constitutes "utopianism," one that distinguishes between those fantasies that *cannot
be done* and those that *could be done if human beings simultaneously got off
the collective tuchus and devoted every ounce of ensuing effort to their realization.*

Paul Goodman, the legendary co-author of the seminal hous-
ing-sciences text *Communitas: Means of Livelihood and Ways of Life*, offers us such a refined conception of Utopia. Instead of lumping all
utopias into the realm of the Fantastical, we must form two catego-
ries: the fantasy utopia and the practical utopia. A practical utopia is
an end that could be reached using currently available resources, if we
could only muster the collective willpower. It is utopian in that it is
beyond the realm traditionally labeled "possible," but it is practical in
that there are no true barriers to its achievement outside of our own
minds. The idea for the youth of Kansas to build a mountain falls
into this category. The only barrier is the youth's willpower! (Perhaps
a few ordinances and statutes regulating large-scale topographical re-

construction projects would stand in the way, but these too are merely products of our will, artificial barriers that we place in our own way.)

One final point must be made, regarding Goodman's suggestion for the Youth of Kansas. Goodman posits that the mountain would create a ski-slope and "manly work." But there is one more benefit, which is the giant hole that would be created from the material excavated for mountain-creation. What community could not find use for an enormous hole? This additional insight into Goodman's plan is known as the Nimni Corollary to the Goodman Doctrine.

Consider how much time is currently spent selecting frozen yoghurt toppings or voting in presidential elections. Now consider what could occur if these wasted energies were diverted, and we simply agreed to forever default to sprinkles and do without a president. At last, we could finally get something done around here.

★   ★   ★   ★

Not only this, but there is plenty of space in which to erect our dream-silos. A surprising quantity of nothing litters the geography of latter-day North America.

The term "fruited plain" is bandied about with such carelessness these days by people with no knowledge of its ramifications that the words threaten to slide into utter meaninglessness. Yet consider what is actually meant by this sainted phrase. In his seminal treatise on the subject, *On the Fruiting of Plains*, Edmund Burke left no doubt as to his position:

> *We do not sufficiently distinguish, in our observations upon language, between a clear plain and a fruited one. These are frequently confounded with each other, though they are in reality extremely different. The former regards the understanding; the latter belongs to the passions. The one is a plain as it is; the latter as it is felt. Now, as there is a moving wave of plains, an impassioned countenance, an agitated gesture, which affect independently of the things about which they are exerted, so there are plains, and certain dispositions of plains, which being peculiarly devoted to passionate geometries, and always used by those who are under the influence of any passion, touch and move us more than those which far more clearly and distinctly express the subject matter.*

Irrespective of the differences *entre* plain-types, then, what can be almost universally agreed upon is that there is an excessive quantity of plains-general in the latter-day United States. Perform an experiment. Take your atlas from its case, and examine briefly our States. Observe the locations of the many cities and towns found within it. Now examine the spaces *in between* these places. You will likely find that there are an awful lot of them. You will, in fact, find that the vast majority of the country is composed of *nothing in particular.*

This reality petrifies us, as a healthy society can hardly be founded upon *nothing in particular.* It must have walls and bars and community pools. All of this Empty Waste means that Man is losing and Nature is in the lead, a state of affairs that should be intolerable to all those who believe in the Civilized Order.

Please, colonize the nothing. Bring your dreams and utopias there. Abolish the abyss, and plug Mother Nature's enormous void. Build a mountain!

LEFT-WING AIRLINES

Let us now consider how to apply Goodmanian logic to a specific social problem, to build a mountain in the air, if you will.

The experience of intercontinental air travel has long been one of immiseration and disrespect. No sooner does the steward bellow "All aboard!" than a sea of bothers begins its briny journey up the traveler's thigh. It is not the food; on the contrary, airline comestibles have always struck us as sensible and efficient, given the circumstances. Nor is it the legroom; we would never hire Procrustes to run our human resources department, but we do believe that those who grouse about spaciousness should consider becoming less spacious.

No, our objection to air travel has always been its rigid class system. Passengers are segregated by income, and given services and comfort in accordance with their economic worth. Those in the upper class cabins are addressed in the respectful "usted" form while those in the steerage berths must content themselves with a lowly "tu." Business class lavatories offer a wide range of scented lotions and themed enemas, while those in the rear trough are required to use leftover SkyMall magazines and laminated instructional safety cards in lieu of toilet tissue. It is a temporary zone in which the material relations of production are acted out to their utmost; it is a mile-high feudal coffin in which each must spend the distance from Milwaukee to Ocala in desolate contemplation of his lowliness.

But when we have offered draft versions of this observation to colleagues during departmental breakfasts, we have been met with nothing but jeers and shunnings. "Yes," says the interlocutor, "your premise that planes are Marxism made manifest is incontrovertibly true. But that is all the more reason for their celebration. After all, is it not far superior to have our economic conditions visible rather than invisible; to have them nakedly exposed at the gate, rather than to unexpectedly disrobe midflight, so to speak? You may be right that a pilot is a feudal remnant, but so are all things, of which the pilot happens only to be the most honest."

The position is a tempting one, though it carries subtle smack-ings of the perspective wielded by certain nameless Slovenian philos-ophers. We have always believed things ought to present themselves as they are, rather than dressing up as something else entirely. We did not hold our tongues when the fruit juice conglomerates conspired to offer watered-down concentrates and dare to market them as the genuine article. Nor did we take kindly to the way in which our 43rd president went off and simply reincarnated himself as the 44th. No, impersonations have never been our specialty.

And yet we are nevertheless forced to condemn the airlines. For lo, though they may do a fine job of sorting us into our classes and being cruel to us, and while this may be an excellent way of being forced to appreciate the true nature of our society, we cannot help but feel as if these airlines compound the outcome by measuring it, to use an apt Heisenbergian insight. We see ourselves thusly, thus we enact the selves as which we have seen ourselves. Someone tells me I am a beekeeper, so I begin to collect bees. A man mistakes me for a carpenter, thus I build him an end-table. Death comes only to those who accept the doctor's conclusion that they have died. And so if I fly business class, I come to feel as if I am in charge of a thing or two. One cannot be in business without being *in business*, as it were. Like-wise, if I am shunted to the cargo hold with the urchins and dross, I experience a remarkable urge to start calling my elders "Guv'na." The poor man, too, receives instruction in his rôle.

One could escape all this bother with ease and grace. Except for brief periods, man has never aspired to a left-wing airline. In fact, we have treated airlines with relative indifference as political entities. *This has been a colossal mistake.* For the classless society begins in the cabin. If the copilots saw themselves as airborne Lenins, even for a day, imagine what might result. We might hasten the restoration of piano players as a central feature of the flight lounge experience. We would certainly foster a unity of passenger purpose, and it would be little surprise if the cabin crew did not at least occasionally burst into the Internationale out of sheer giddy revolutionary intoxication.

They say an aircraft with two left wings cannot fly. But to this we say "How can you know for certain, and it isn't it worth the risk to find out?"

SIMPLE IMPROVEMENTS FOR THE MODERN-DAY MOTORCAR[79]

This work is, at its very pulsating core, a guide to innovative new methods for edifying the human experience. And it is a fact that humans experience motorcars more than perhaps anything else. We therefore propose to use this Chapter to revolutionize wheeled transit, having already taken on its airborne counterpart.

Take the modern American automobile or jalopy. It looks something like this, yes?:

A well-thought out device, certainly, but it is also a restless one. It suffers from a number of internal tensions which cannot be overcome by willpower alone. Everything Marx said of capitalism applies equally well to the Volkswagen. Furthermore, its emphasis on the metallic over the organic cannot be excused by appeals to efficiency or speed.

---

79 Note: this chapter was written before our views on horses had properly solidified. It thus contains a series of positive remarks on transportation mares to which we can no longer in good conscience adhere. It is nevertheless included in this work as a primary document, evidencing the steady evolution of our thought.

We therefore now ask you to consider what would occur if this contraption were to be modified in the following manner:

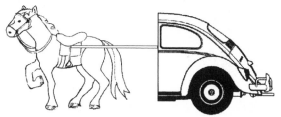

We humbly suggest that this is a far superior design for a transportation-machine. We have removed the contradictions as well as two of the "wheels," allowing for greater ease of movement as well as wind-flow.

Our new design is born from a universal principle: *Everyone like horses.* After all, remember the old advertising slogan, promising "Horses with their breakfast, horses in their tea. Horses wherever they go, and an evening horse just before lying down to sleep." *Horses are to mankind what whales were to Melville.*[80]

Some, however, may neigh-say our improvement, and justifiably register alarm at the paucity of airbags present in the Car of The Future. We have noted this criticism and accounted for it. As careful observers of the blueprint will note, airbags have been cleverly replaced with an enormous horse's ass. We believe that the improved fertilization of America's streets will more than compensate for the reduced safety of her automobile travelers.

If you or your loved ones would like to find out more about horses, we recommend seeking out the following horse-centric printed works:

- Irwin, Chris. *Horses Don't Lie: What Horses Teach Us About Our Natural Capacity for Awareness, Confidence, Courage, and Trust.* Marlow & Co., 2001. $14.95.
- Kohanov, Linda. *The Tao of Equus: A Woman's Journey of Healing and Transformation Through the Way of the Horse.* New World Library, 2001. $24.95.
- Parelli, Pat. *Natural Horse-Man-Ship: Six Keys to a Natu-*

---

80   Aphorism #19. Again, please take into account that this entire portion of the book has since been disowned.

ral Horse-Human Relationship. Western Horseman Press, 2003. $17.95.

- Pelicano, Rick. *Bombproof Your Horse.* Trafalgar Square Books, 2004. $24.95.

(Please note: several of these works are printed on horse-based paper.)

But there is more. Automobile anthropomorphism is another must, if we are ever as a civilization to achieve the levels of car-cuteness depicted in our literature. No two speaking-cars have ever attacked one another, which led us to formulate the Anthropomorphic Peace Theory framework for examining traffic safety issues.

Yet perhaps it is time to reconsider our position more completely. "Here, you," says the intoxicated peer-reviewer. "Why are you trying to repair the motorcar instead of throwing a grenade into it entirely? After all, is not the most extreme measure usually the best?" "Yes," we might reply, "but sometimes the most extreme measure is not, in fact, the best. Furthermore, one tends to get less far throwing a grenade into one's car than one gets by repairing it."

Still, there could be a point buried amid the reviewer's violent idiocy. Can the motorcar ultimately be redeemed? Perhaps not. Why, just take a gool's gander at its guts:

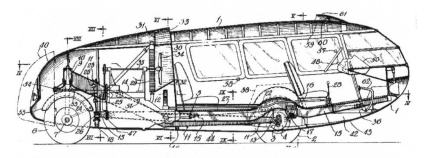

No human-material object more conspicuously and proudly violates our prior aphorism that *nothing must look and act like a fish save an actual fish*.[81] While this pithy phrase has been known to result in a stern look or a vicious pounding when used as a pick-up line in a canasta club, its validity in matters vehicular is somewhat unquestionable.

---

81   Aphorism #20.

None of this is to suggest that the wheel is obsolete, though surely it is. We do feel, however, as if the time has come for motion-science to proceed to the next level. Just as the circle improved upon the rounded square before it, and the rounded square upon the human leg, *something must come next.*

It is time then, to think clearly about what the subsequent phase in transit innovation will be. There are those who believe that when cars can drive themselves, they will finally be happy. There are some who think it is time for a disastrous attempt at a space elevator or two. Finally, there are also those who believe nothing short of *pod-based transit* will satisfy a hungry Zeitgeist. Under this proposal, all transit would not occur in privatized motor vehicles, but would be a series of individual public pods on overhead rails, each of which is designed to carry a single family.

Doubtlessly, the pod the simplest and most elegant of the Forms. Let us consider what Ms. Rand herself had to say on the subject:

> *"If one wishes to advocate a free society—that is, capitalism—one must realize that its indispensable foundation is the principle of the pod. [And] if one wishes to immanentize the pod, one must realize that capitalism is the only system that can uphold and protect [it]. And if one wishes to gauge the relationship of the pod to the goals of today's intellectuals, one may gauge it by the fact that the concept of the pod is evaded, distorted, perverted and seldom discussed, most conspicuously seldom by the so-called conservatives."*
> -Ayn Rand, "Man's Pods." From <u>The Virtue of Self-ishness</u> (1965)

But as avid consumers of mid-century speculative fiction, we cannot escape a trepidation when the word "pod" is mentioned. As anyone who has seen film adaptations of *The Pod People*, *The Cantaloupe Women*, *City of Pods*, *The Pod Horror from Earth*, or *Attack of the Pod-Based Transit System* will remember, nothing good comes when humans dabble in the concept of enveloping themselves in pods, however innocent their intent.

What next, then? The cable-car? The jolly-trolley? Those who think in these terms have been confined in small rooms. We set our ambitions a wee sliver higher, aiming for nothing less than the *transcendence of movement itself.*

If it is considered rationally and without blinkerment or prejudice, movement becomes a highly illogical phenomenon. Why do we, the human race, move? The question can first be answered empirically. We move to go and see our loved ones. We move to get a bit of exercise. We move to grab another root-beer from the coolerator. But we do not move for *movement's sake.* Movement is nothing more than a means to a series of highly enjoyable and sometimes perverted ends. It has no intrinsic worth. In fact, too much of it too quickly makes us nauseous, as those who have found themselves vomiting an afternoon's worth of funnel cakes aboard a state fair Tumble-Bug ride surely know too well.

No, if it were left up to us, we would sit right here and rot in bliss. But when one rots, so does one's mind, and as that beloved-yet-tyrannical elementary-school poster insisted on reminding us each day, "A mind is a terrible thing."

How to eliminate movement, then? Naturally, we might start by bringing the coolerator closer, or even condensing all the functions of the contemporary home into a single palm-sized unit. Yet what if one's relatives live in Cheboygan? Certainly a visit is out of the question, yet as that beloved-yet-tyrannical elementary-school poster insisted on reminding us each day, "A family is a terrible thing to let starve to death in the tundratic wasteland of northernmost Michigan."

But perhaps we have spent too much time in the car. Let us alight, and wander into the buildings, to see how they, too, might be improved through an application of Academic Insight.

BRINGING GEODESIC STRUCTURES INTO COMMON USE

As has now been sufficiently documented by both textbooks and teleplays alike,[82] "the genius contribution of Fuller was his discovery that the omnitriangular frame of the geodesic sphere gave the strongest

---

82 *Buckminster Fuller: The Story of a Discreditable Crank*, BBC Television, Innovators & Eccentrics Series (1986).

possible structure per weight of materials employed."[83] It has surprised many of us that, given the magnitude of this fact, Fuller-worshipping death-cults have not sprung up across the land. (Yes, there are reasons to believe death cults are bad idea,[84] but this *in itself* does not make their absence explicable.)

Still, however, there is reason for hope. For while Fuller may no longer be among us, pooping out geodesic domes at every stop on the lecture circuit, and while the domes that were initially excreted have now slipped into decay and ruin, a new wave of dome-conscious youngsters is set to take its place at the forefront of the construction and development industries.

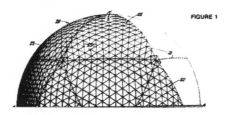

FIGURE 1

Typically, of course, favoritism has no place among science-groupies, but we believe that the prioritization of the geodesic over the merely desic is supported by both rationality and conjecture. While each individual dome, with its individual dome-personality, can only be built by mortals, its form is eternal and universal, meaning that the dome approaches the status of a geometric deity, if such a thing can be said to be. We are of the firm belief that if all buildings were dome-shaped,[85] warfare between peoples would not only cease, but would become unthinkable.[86]

In addition to our dome-advocacy on a purely constructivist level, we are firm supporters of the proposed Dome Amendment to the

---

83   Ibid.

84   In saying this, we retreat somewhat from the position espoused in our early pamphlets, notably "Domes and Dissent: Against 'Live and Let Live' for Anti-Geodesic Counter-revolutionaries."

85   The statement "all" buildings is, of course, academic hyperbole. Exceptions would naturally have to be made for stand-alone lavatories and emergency telephone booths, the shaping of which into domes has proven both costly and unsightly.

86   We will leave it to others to furnish more in-depth justifications for what we call our Geodesic Peace Theory.

United States Constitution, which would provide for the doming of all objects within the national boundaries, in a similar fashion to the point-based law governing Oblio's Pointed Village. For our Fuller has found the noblest of the forms, and has given us the tools we need to reshape all we hold dear in its (and his) image. For too long we have contented ourselves with rectangular or oblong postage stamps and billboards, unaware of their longing for greater shapes.

Long live Buckminster Fuller. Viva la dome.

THE HABITABLE CAPSULE

For all the ideological posturing conducted by this most recent generation of "domestic engineers" and "religious scientists," we have yet to create a functional alternative to the modern-day housing unit. We continue to live, for the most part, in a planet-sized uninhabitable hovel. But why should this in fact necessarily have to be the case, given that the solution is pouring us right in the tea? We speak, as readers will already no doubt have discerned, of the Habitable Capsule.

Discontentment among the more fashionable sectors of the materials-science community has engendered general skepticism about the possibilities for habitable capsules. Yet surely in light of our present housing crisis, capsule resistance is not only unhealthy, but potentially lethal. *To disrespect the capsule is to disrespect one another.*[87]

The capsule is the most universal of human-created forms. Structural engineers have long sought to replicate the capsules found in the nuts and pods of nature with artificial materials (polystyrene, lanolin, pleather, etc.) It is the central tragedy of the Holocene Epoch that those who have succeeded in this endeavor have turned their attention to the manufacturing of soap dishes and pharmaceuticals rather than the solving the true problems of the day.

If such men and women of genius were to instead focus on the creation of such "habitable capsules" we may find our living quarters revolutionized and our cities salvaged. After all, modern man, "who no longer dresses in historical garments but wears modern clothes, also needs a modern home appropriate to him and his time, equipped with all the modern devices of daily use."[88]

87 Aphorism #21.

88 Source misplaced.

Above, please bear witness to the construction of one of the proposed habitable capsules. As can plainly be observed, the numerous small individuals building the capsule find merriment and fulfillment in their work, and as the capsule comes together, it finds a dual function as both a solution to the population crisis in the summertime and a steamy indoor jacuzzi in the wintertime.

To our friends and the mildest of our nemeses, we therefore offer this modest prediction: the future of Mankind is to be not only capsule-*based*, but capsule-*derived*, as all life is at the core. As the Japanese sleep in tubes, so will all from the Bangladeshi to the Statesman sleep in the habitable capsule.

New and Better Housing Units

But we are dwelling too much on solutions without looking at problems. It is necessary to elucidate the precise reasons why the Modern House has become an impossible place.

★   ★   ★   ★

We shall diagnose the problem by sight, from a distance.[89] With several notable exceptions, today's housing-contraptions still look roughly like this:

---

89 For the development of this technique, we are grateful to Senator Bill Frist (R-TN)

A humbling and tasteful dwelling, to be sure, but one which not only ignores but stands in broad defiance of the full range of human potentialities. It is not in that category of human structures "that by their individual character give form and continuity to the life that goes on within them," as Lewis Mumford so pleasingly put it. It may serve as an adequate vessel for the preparation of a meat sandwich or the intake of Beer and Television, but it is not *ennobling* of the human spirit. It resigns us to its flatness and its give-upitude. It is a building which has become so thoroughly disillusioned with *purpose* as to negate the very act of *livingness*.

Furthermore, the house in all of its stupefaction falls into the trap laid out in the early portions of our text: it allows itself to become *unitary*. No matter how well these units are stacked upon one another, they fail to conduct the *clustering of labors* necessary for the maximization of human productive power. If all things do not occur in the same exact place, what is the purpose of their occurring at all?

Now, say that instead of the above design, the housing-unit were reimagined as a module-based system along the lines depicted below:

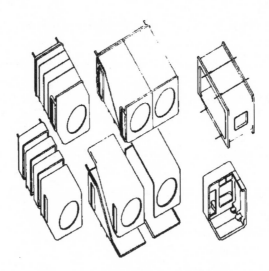

A system of interlocking component-housing-parts such as this would, we posit, not only *save* certain kinds of labor but eliminate the need for them entirely, as the circular or "porthole" window has been shown to carry *maximum dynamic efficiency of purpose*. Leaving little room for "sofas" or other indulgent accessories as it does, our plan may be derided by comfort-theorists as overly spartan or unnecessarily egalitarian. We ask these mal-informed critics, however, to rebrowse their copies of Leibniz's *De Arte Combinatoria*, in which the great Saxon states clearly and without qualification that *la vie de la gousse est la vie noble.* ("the pod-based life is the noble life.") Those critics who follow this suggestion will, we posit, be left utterly unsure of how to reply.

The Window as Door

One of the unseen advantages of our proposed housing-device is its subversion of traditional exit-entry rôles. One of the more insidious dogmas of architecture both historically and presently has been the door-window dichotomy, in which the passage of light and the passage of bodies into and out of structures are considered as separate and non-intersecting functionalities. We consider this to be a grossly inefficient doctrine to cling to, and have been struggling throughout the course of our adventures in the field to fathom the reasons for its persistence. Despite some effort, we have thus far fathomed none.

We therefore argue for, and present the groundwork for, a merging of these two core architectural purposes, and propose that no "window" that cannot be passed through with equal ease to the modern-day "door" be spared from the torches and pitchforks of the New Socio-Architectural Revolution.

If the maximization of human energy (measured in joules) and freedom (measured in freedomlets or *joules de liberté*) is truly the purpose of spatial planning, then there can be no justification for any conscious design-decision which reduces rather than maximizes penetrability of a structure.

Furthermore, we need hardly mention the advantages of our philosophical-architectural method in the reduction of needless Death By Fire, as well as in the creation of communal communication. That is to say, if all of the walls are holes instead, one can both easily flee a conflagration and yodel hello to one's meter man. All of this speaks to the gross nearsightedness of seven centuries of architectural "progress" and the supreme superiority of the "window as door" outlook on both design and life itself.

Tubes Like Hamsters

If we move even further beyond the Window As Door framework, however, we arrive not at a mere transition but at sheer transcendence. What next step could lie beyond the elimination of the door? Tubes like hamsters. The Japanese have pioneered so-called "difficult" housing, which aims to fortify and nourish the occupant by providing physical challenges to intra-habitational movement (topsy-turvy krazy-stairs and blindingly bright, deliriously clashing wall-colors, for instance). We believe that using hamster tubes as the primary mode of transit within the dwelling will both strengthen the health of the liver[90] generally and increase the interconnectedness of rooms within the habitation. Doors are barriers and blockages. They are clogs[91] in the arteries of the living, breathing home. A home is an organism; it is the Oyster of Man. When we put barriers within it, we irrationally compartmentalize our activities. When we tubularize it, we maintain distinctions between living-functions without severing the flow of action that bleeds from room to room and pastime to pastime.

Handles

When considered thoroughly, as some things ought to be considered, the fact that we have both thousands of empty homes and thousands of homeless persons is a remarkable non-success. How have we allowed the Incentivists and Economists to bar the route from A to B (where A is our current state and B is Filling Every Empty House With A Person Who Has None)?

We propose increasing the efficiency of the aforementioned channel; *no human who wishes to have an inhabitable dwelling shall be denied one.* To this end, we recommend the mass increase of handles on recently-constructed homes. Aside from an irrational attachment to notions of "just desserts," the central obstacle between Us and Complete Homedness is the portability of the present-day home, which presently stands at Zero. The very fact that a separate category exists for "mobile" homes shows just how far from mainstream the idea of portability has become.

90   as in one who lives, not the foul-looking digestive organ

91   as in Blockages, not the charming Dutch footwear

If handles were attached to each home, and they could be transported at the leisure of the wearer, future housing bubbles may not Pop but may instead float gloriously on the breeze, growing ever-larger without the risk of being impaled on nearby steeples.

Naturally, we support expanding the portability of *all* manufactured items, but we must start with the home. After all, without a home, a person is but a wandering fleshy tumbleweed.

Gravity Defiance

As the vast majority of American plains are fruited yet lacking in utopian micro-towns,[92] so the vast majority of the space within the contemporary home remains devoid of human activity. Consider a "living" room of 15 by 12 feet, with eight-foot ceilings. It contains, according to our best approximations, 1440 cubic feet of livable space. Yet the contemporary human being (dimensionally averaging roughly 1 x 2.5 x 6 feet) takes up only 15 of these cubic feet. The human could be stretched the length of the room with the entire width to spare. 1425 spare cubic feet! Consider their possibilities!

But we do not propose to make the living room *smaller*. We propose to make human activity *larger*. One of the primary reasons so much space remains unused is the irrational preference of feet for the floor or ground. While their motives (stability and ease of movement) may be respectable, they do a disservice to the being they transport, by drastically cutting the scope of potential intra-room destinations.

We believe that through the addition of flotation to humans within the home, our space will be better used and our potentials better fulfilled. If one cannot touch the ceiling, why have it at all?

Elimination of Furniture

There is, simply put, no reason why furniture should be distinct from the home itself. After all, if a home is to be a living-unit, all parts of it should serve the user usefully, and it should need no additions or modifications in order to fulfill its charge.

Take the bed. Why must we sleep two-and-a-half feet above the floor? Why can our floors themselves not be soft and mattressy,

---

92   *"Stalks of wheat grow best in soil fertilized by oppression."* - Aphorism #22. Note: This is not strictly an aphorism *as such*, but a quote from Joseph Stalin.

equipped with all of the comforts necessary to slumber us? The same logic applies with equal rigor to the sofa or *chaise longue*. Must they stand separate from the floor, or can the floor be transformed into an enormous beanbag? Televisions need not have cabinets when we have perfectly good walls to build them into, although we also support the elimination of both the television and the wall.

By incorporating all furniture into the house itself, we increase the occupants' identification with their home. It is not a storage unit for personal items, but *is* the personal item, and cannot be "moved from." Furthermore, we eliminate unsightly wires, as well as the *sous*-furniture crevices that insist on compiling unsightly collections of dust and "spiders."

URBAN PLANNING AND YOU

But there's more. Houses, when woven together, become a quilt. The city is comprised of not *one* structure but *many*, and this aggregated metastructure has properties of its own. The care and taming of these properties is called *urban planning*, and while most people are too busy writing scholarly articles to engage in it or understand it properly, it is in fact a pastime of some considerable consequence.

In fact, we doubt that you have the slightest *clue* how much the incredibly tedious and mundane field of urban planning affects you. Each day, the majority of your life decisions and experiences are structurally determined for you by the actions of some monocled technocrat from dozens of years prior.

Do not take it from us, however. Take it from these experts, who have written a book published by Random House, a prominent and well-respected publishing company:

> *"The works of engineering and architecture and town plan are the heaviest and biggest part of what we experience... the background of the physical plant and the foreground of human activity are profoundly and intimately dependent on one another. Laymen do not realize how deep and subtle this connection is."*
> -Paul & Percival Goodman,
> Communitas: Means of Livelihood and Ways of Life
> (revised edition, 1960)

When Mr. and Mr. Goodman say "Laymen," they are speaking of You. *You* do not realize the extent to which designs of the physical world affect you. Consider the highway on-ramp. If it stops in your neighborhood, you may ride the highway to freedom and beyond. If it does not, you must travel a million miles of muddy back-ways and slimy avenues in order to access the United States's primary vehicular arteries. Let's say you desire a job in the next town over. All very well, if Dr. On-Ramp has decided to bless you with passage. If not, however, godspeed in your five-hour daily commute. Hope you've got a well-stocked audiobook collection!

Who decides whether you receive an on-ramp or not? Is the person female? Do you know her name? How can you convince her? Do you know the answer to any of these questions? Of course you don't. Nobody does. The schemes of the Planner are shrouded in a haze of witchcraft and enigmaticism. Yet minute decisions over where to place This or That may have a profound effect on the substance of living.

In fact, What We Build may have a far more direct effect on our behavior than any piece of Law or dictate of Some Silly Government. Consider the speed bump. It is not a law. It is an object. Yet as the law imposes its will upon us, so does the speed bump. It does not threaten to haul you off to the pokey if you disobey it. It merely threatens to ruin the underside of your 1993 Pontiac Sunfire, wrecking both your suspension and your dignity as it sits giving you a look of mocking nonchalance while you nurse your bruises from within the beaten husk of your former automobile, glaring at the bump's undamaged humpitude and wondering how something so innocent-looking could emerge so unscathed from your 40mph act of quixotic defiance.[93]

You are powerless against Architecture. It will tell you what to do and where to go, and you will Obey. Every building is a simplified version of the Hedge-Maze. Have you ever attempted to rebel against a Hedge-Maze, painstakingly prying branches apart and crawling through the tiny subversive hole you have made in the bushes? When you came out the other side, you had twigs in all of your orifices, didn't you? The same problem occurs when attempting to walk through walls.

---

[93]   Only the "quixotic" act is ever worth taking, a sentiment confirmed by Aphorism #23: *"Tilting is not the only action that need be directed at windmills"*

Our actions are controlled by our spaces. Only those with control over spaces have control over life. This is why True Revolutions will seize control not over the Means of Production, for these are static, but over the Means of Design, which hatch the blueprints for the Means Of Production. The coup should be directed not at the President, but at the Planning Board and the Development Council. Otherwise, when you have become the Imperial Potentate of your glorious new state, you will still find yourself being funneled through life like a ping-pong ball, subject to the Founders' floorplans if not their statutes.

★   ★   ★   ★

This also means, of course, that it is time for the People's Zoning Code. For too long, the Zoning Code has been the purview of busybodies and left-brainers. Every time we open the pages of *Municipal Ordinance Quarterly*, some citybound nitwit is enforcing a new scheme for divvying up the humanosphere into a thousand little thimbles.

This is all of our faults. If the somnambulance of city planning meetings was resisted instead of avoided, somewhere we might start to get. It is not necessary for zoning to be a councilman's errand-- it could be all of ours.

Let us go into some specifics, then, as to the provisions and practices that one might see once local planning is done correctly, and pre-existing municipal ideologies have found themselves rezoned.

LAZY RIVERS FROM HERE TO THERE

Nothing brings more joy to the human carcass than to spend an afternoon meandering down an artificial stream in a bright orange inner tube. There is a reason all water parks now come standard with lazy rivers, and that the cultural imaginary has been so thoroughly soaked by these gentle canals, from that early Egyptian hieroglyph of a inner tube between two lines[94] to the 80's power ballad "Is This Love That I Feel? (More Than A Lazy River)" which we were never fond of during its heyday, but whose chart performance we nevertheless cannot help but stand in moderate admiration of.

Yet why confine the lazy river to the water park? Why not it spill out into the *national* park, or wander down Main Street? Why not incorporate it into a broader culture of slow-moving transit canals, littered with gondolas and inflatables? Why is it that I, as a human, cannot simply set foot outside my door and board the next inner tube to Tulsa, *should I wish to go there?* The confinement of the lazy river to institutions that already have plenty of water to go around seems to us the height of redundant misallocation.

It would be hard to deny that a university would be enriched if a slow-moving canal could be taken from the dormitorium to the seminar, or that the office would be transformed for the much better if one's work product could be floated over to the supervisor via tube-tray, so that one need not face her wrath in person.

The suggestion, then, is that the lazy river be adopted as a standard means of conveyance, as basic as the roadway, tramline, and sidepath. Wherever there is a pavement, a lazy river must be placed next-door in parallel, so that *consumer choice* is the operative determinant of transit outcomes. If I am not given the *option* to flee the police in an inner tube, how can we conclude that my decision to use an automobile was my foremost preference?

Let us briefly speculate on the possibilities. First, consider your day as it is presently. You thump from bed to floor, fasten your denticles, gulp your Liquid Breakfast, enter the lavatory to unleash a cumbersome stool, and head out the door to work (work is a factory). The majority of the day will be spent contemplating the void.

Now let us imagine things slightly differently. All of the above is the same, but instead of getting on the subwaybus or railtube after leaving the house, you are met at the threshold by the family gondola, careening at four miles per hour down the local lazy river. "Good morning, Clemence!" you shout to the gondolier. "Good Morning, Mr. Braemis!" the gondolier replies. Clambering aboard and lodging your briefcase under the seat, you are gently rowed through the city, gazing in wonder at the as you meander through the canals to work. "How fortunate I am to live in a city where canals are not the exception, but the rule!" you muse to yourself, sipping smooth Viennese tea from a thermos as the scenery says hello. The gondolier is a stimulating conversationalist, and speaks to you of time, geology, and games. You and he have a quick verbal chess match by calling out moves to one another. (But he also knows when you wish to be left in thought, and a silence between you always carries understanding rather than discomfort.)

With this fantasy having been carefully exposited, then, here is a discussion question. Why must we live in the one reality rather than the other? Is it not perfectly possible, in a time of gross technological intervention into everyday life, to envisage a culture of canals? *Is there anything stopping us*, save the vast expenditure of resources, a total lack of commitment from the populace, and a dessicated political apparatus? Lazy rivers are the way forward, and if our morning commutes are to be not just erotic, but veritably zesty, we must travel down them with haste.

It may be replied: *"All of this is very well, but how do the gondoliers get to work?"* The economist might find this question infantile. "The market, it shall provide," he caws. But it is the economist himself who is infantile. *The gondoliers get to work on gondolas*, of course. Markets are all very well, but heavy coins would sink in a canal. He who fails to recognize this is naive, and we posit that the aforementioned Economists may have been in the academy too long to be realistic.

It will be objected by pessimists that the United States already has at least one gigantic lazy river, her storied Mississip{pi}. Yet this river, like most, is disgusting, and this is its chief drawback. The lazy rivers we propose are both chlorinated and carbonated, detoxifying the flesh while tickling it gently with bubbles.

It will also be argued that if people rode lazy rivers all the time, they would frequently get wet. We take the point. Nobody has been more skeptical of the value of water than we.[95] But we would pose our own question in response: Does getting wet generally cause lasting pain or damage? Unless one is a piece of sensitive electronic equipment, the existence of which we oppose to begin with, the answer is no.

We have found ourselves repeatedly horrified by the indifference with which our proposal has been met. Each time we encounter a pro-

---

[95] Was it not we who originated the scrawling of that now-classic piece of graffiti "Water: "They call it the Universal Solvent, but has it ever solved a tabletop parlor-quiz or quelled an insurrection? Hardly!"? It was. This footnote should in no way be taken to constitute an admission of vandalism.

letarian, whether he is fixing our garbage disposal or fixing our porch light, we detain him in order to outline the fundamentals. Yet instead of the glow of intoxified enthusiasm with which we expect his face to be overcome, we are simply given gruff indications that time spent examining our diagrams will be added to the service charge. The prison of ideology keeps the public from recognizing its own good.

That squandering of our foresight will hopefully be corrected upon the publication of this book. Hopefully it will be recognized that, aside from those with engineering expertise, no two experts can pontificate with greater authority on the subject than ourselves, considering our status as the twin co-founders of The Miles Davis Celebrity Whitewater Expo, an event driven by the principle that certain kinds of rafting are in themselves a jazz.

And yet, in order to justify traveling lazy rivers to get from building to building, there must be buildings worth going to in the first place...

Much Smaller Buildings

It was when we found ourselves riding an elevator to get to another elevator that we finally realized humankind had lost its architectural marbles. As we passed floor after floor of needlessly stacked layers of office, we wondered why man ever felt it was necessary to "scrape" the sky. You don't try and chisel the moon to bits, why should one go scraping pieces of the sky off? Won't that make it look patchy?

It will. Today our buildings are far too large, and the reasons for their being so largely fail to grow beyond the penile. We do not know who it was who first came up with the notion that buildings ought to be tall, but we suggest that the failure of this man's closest friends to execute him when they had the chance stands as one of history's greatest blunders. Not since Mr. Schicklgruber decided that tonight would be a pleasant night to put his privates someplace warm has such a trivial error resulted in such mass misfortune.

The problem of tall buildings is difficult to get to the bottom of. Why erect a thing that one must elevate oneself to surmount, instead of one that one could stroll to? The architects who conspired to upwardly elongate our structures, to arouse them skyward, cannot possibly have considered this question. After all, each skyscraper could gently be laid on its side to produce a perfectly serviceable longhouse. The failure of designers to pursue this course can only speak to a lack of awareness of the longhouse's storied history and highly regarded geometry.[96] Now, this ignorance-of-longhouse itself may serve as an indictment of the American system of schooling, but nobody who has read this work in its entirety can accuse us of neglecting to offer useful fixes for the schools as well.

But perhaps we do not face a binary dichotomy between long and tall. Could buildings not possibly be small? "Getting small" is no longer the taboo it was during its 1970's heyday, even if the "small is beautiful" gang have been exposed as dreamers and longhairs. It might yet be possible to build a workplace that is less than seventy stories tall.

The proposal may seem radical, even desperate. But we assure our reader that the task, while arduous, is not beyond human civilization's sum total capacity. Our skies may yet escape unscraped.

Yet Non-Religious Cathedralism

But there is a small beetle of worry floating in the lye of progress. If we confine our buildings to certain shapes and sizes, do we not forgo the monumental and breathtaking? Everyone can admire a Lego Notre Dame, but would we be happy to have it replace the original? Some would say not.

Thus it is vital to balance two different oysters in one's hand at the same time: we ought to mostly inhabit tiny cottages, but there must nevertheless be pockets of grandiose splendor dotted across the fourscape. Fortunately, others than ourselves have come up with the solution.

On this we defer to Krier, who emitted the following formula: *limit all buildings to four stories, but allow them to be any number of feet.* This way, we may still have our vast edifices and erections, but only when we are absolutely sure we need them, instead of having them pop up all over the place when we least expect them. No more elevators to skydecks to get to further elevators. A lobby, an upper viewing deck, a

---

96   See Michael Johnson, *Iroquois: People of the Longhouse* (Firefly Books, 2013) and Bruce LaFontaine, *Wigwams, Longhouses, and Other Native American Dwellings* (Dover History Coloring Book Series, 2014).

penthouse restaurant, and an observation platform are fully sufficient to comprise the sum total of any building's layers. Who in their life has enjoyed living in an apartment building higher than four stories? And at the same time, who has not enjoyed a high ceiling or two? Our program welds the best of both delights.

Furthermore, it does not entail the elimination of the Room-With-A-View. The Room-With-A-View may still exist, but must simply sit atop a vast hollow hundred-foot chamber, *which will itself be enjoyable to stand in*. And there need be no more Rooms-With-Slightly-Less-Of-A-View; one is either atop or at bottom, with no in betweens.

Secular cathedrals and small, friendly brownstones, then: these are the chief foundational elements of the New American Urbanscape.

THE REINTRODUCTION OF PUBLIC BATHS

We are concerned with the Schisms and Fractures that increasingly dot the flesh of the nation like the buboes of old. Not only do we cling fiercely to the notion that a healthy society requires General Goodwill, but we believe much of the present conflict stems from misunderstanding rather than malice. Having tested the hypothesis between ourselves, we happen to know that if all persons simply sat garmentless in a circle and discussed those socio-political matters presently bothering them, inter-personal conflict would decrease to an almost negligible minimum.

Knowing this as we do, the following is hereby proposed: The Public Bath shall return to American Public Life, rectifying the century-long injustice wrought by the Sanitation Standards Act of 1902. The monopolization of the bathhouse concept by the sexually adventurous has sapped the democratic energy and legitimate social function served by the institution in its historical Romanesque form.

Therefore: A communal bath shall be placed along the main road or avenue of each town with over 5,000 residents, while in rural areas one shall be placed at the exact center-point of a geometric circle encompassing the residences of 5,000 persons. Regular bus and electric-monorail services shall transport persons living at a distance of four miles or more from the nearest bathhouse to their closest location. Furthermore, all municipal, state, and federal elections are to be conducted within the bathhouse, with ballots as confidential as the contents of the bathgoers' vestments (i.e., not at all).

This radical re-centering of public life, and the alignment of the dual duties of *citizenship* and *cleanliness* would both energize democracy at the local level and gradually eliminate the discomfort Americans tend to feel when confronted with one another's private parts.

The rigid line between public and private has caused our society to splinter and compartmentalize. By bringing the private(s) into the public, we will revive the grand Turkish traditions of communal deliberation and collective bathing.

THERE IS NO SHAME IN RITUAL (EXCEPT FOR RITUAL SHAMING)

But collective nude bathing is just one way of increasing national togetherness. The gulley of the point flows into a wider murky stream, namely the necessity of ritual in orienting the human day of the daily human.

Aside from those which involve piercing one's eyestalk or sacrificing a street vendor, rituals can be a healthy way of orienting the body in time. It is true that the ugly history of tarring-and-cheesing the tax-collector, or conducting musical lynchings, should give us pause before we voluntarily sweep ourselves up in a pleasant mass hysteria. But for every ritual in which portions of an infant's penis are sliced off and discarded, there is another in which toast is eaten with the homeless, or pop songs are bellowed in harmony as railroad spikes are driven.

Our current rituals are embarrassingly impoverished. Each night we send up the same fireworks while wearing the same stale dresses. Yet imagine a ritualistic culture in which a barn was erected together every fortnight, or a Mount Rushmore was destroyed every hour on the hour.[97] If little boys were not given bar mitzvahs, but were

---

97 Is Mount Rushmore a defensible edifice? Yes, there are times when each of us daydreams lustily of a time in which our own têtes will be carved into a rock face. But this in itself is insufficient justification; after all, sometimes I aspire to be my own Godzilla, eating New Yorkers from the treetops, but this does not mean I should be given a grant from the Ford Foundation in order to do so. It is difficult to conceive of a more comically crass affront to the continent's native inhabitants than the enormous faces of four presidents carved into sacred land. Our treasured mount is a ultimately a gaudy slice of geological kitsch, and the symbolic value of its destruction through the use of heavy explosives would be incalculable.

sent over small waterfalls in a masculinity barrel, would not adulthood seem that much more of an achievement?

We ourselves live entirely in ritual, and recommend others do the same. We have our ritual coffees, ritual danishes, ritual finger baths, ritual harangues, and ritual teeth-brushings. Each part of our today is a synchronized communion with Time and Cultural Practice. This does not mean there are no surprises; our recent addition of the Ritual Search for a New Adjunct Position came entirely unexpectedly. But it does mean that our activities have value solely to the extent that they are conducted ritualistically. *There may be an I in Team, but there is no shame in ritual.*[98]

INTERNATIONAL RENAISSANCE FAIRE

Speaking of crucial rituals, we would briefly note that we have historically enjoyed ourselves at annual Renaissance Faires, and feel as if they will likely form a core part of the future society. To this proposal, some may cower in dread at what sounds to them like the world's longest future, but no conceivable social arrangement can make room for all possible opinions.

PERMANENTLY DISPOSING OF SPORTING EVENTS

Yet not all costumed ritual hysterias are equally nourishing. We have long felt sporting events to be far more symptom than disease, and thus recommend their immediate abolition. No utopian end has yet been served by the tossing of a "pigskin" across a "court," and it is a reasonable supposition that *none ever will be*. Indeed, the very conception of a sport itself is designed purely to satiate our basest and most purely elfin impulses, and those who voluntarily subject themselves to "the thrill of the game" (*l'esprit de l'escalier*) are willingly signing themselves into bondage and enslavement through a binding blood-contract.

Augmenting this dissatisfaction on our part is an equally potent dislike of cheerleaders. Those whom we have met, we have found to be universally vapid and unpleasant, with a sickening saccharine faux-bubbliness belying the cruelty and backstabbery that yearns to spring forth from their dark, spiteful hearts.

[98] Aphorism #24.

Therefore, the eternal eradication of bodily "sports" is an essential condition for the continuation of human progress. The stadiums must be turned into drive-in hospitals, and the courts, pitches, and fields into summertime leisure-palaces or practice grounds for state-sponsored Morale Improvement Marching Bands. Uniforms will be ground up for feed or donated to one of the seedier children's museums

A. Structure and Infrastructure

Education, Part I:
Giving Newborns Pipes

We have not dealt substantively with the issue of child-rearing in our work thus far, as we generally entertain a pluralistic and tolerant view of parenting (except as regards our own offspring, who are instructed to write daily reflections on the *Blueprints* and forbidden from the use or reading of poetry). However, we are obligated to briefly delve into the fetid swamp that is The Young, as our program for the readjustment of society must necessarily confront the sour fact that Children are the Seedlings of Tomorrow's Corpse-Flowers. Therefore, those uneasy parents who have been frustrated by our text thus far, tentatively gobbling up our thoughts but burning with the question "But What Do I Do With My Child-Persons?" may now reach a state of peace, as we instruct you on *precisely* what to do with your child-person over the course of six extended mini-encyclicals corresponding (roughly) to the six stages of child-development.

★　　★　　★　　★

The study of infancy is only in its young adulthood. But there are proven ways in which the life of a baby, generally an alarming experience, can easily be mitigated. It is not impossible to simultaneously be a baby and enjoy oneself. Of course, were it possible to bypass the early post-natal years entirely, to emerge fully formed from the birth canal already bearded and in tweed, it would be universally agreed to leapfrog what is known among developmental psychologists as the *peeandeat* period. In the absence of the ideal, however, we must make compromises.

True, one cannot be born with elbow patches and waistcoat already sown on. But for every thing, there must necessarily be a *next best thing*. As it is, regarding infancy, we generally favor a pipe-based reform program. About this, several points must be made in rapid succession:

◆ It is our firm conviction that if newborns were given smoking-pipes, they would achieve an unexpected degree of sophistication in an unprecedented amount of time.

◆ It is said that the clothes make the man, but this maxim is only correct if the man is simultaneously smoking a pipe. The pipe makes the man, and the clothes are but a foul deception.

◆ Since Magritte was unwilling to tell us what a pipe *was*, we hereby offer a definition: no tubular object can be said to be a pipe unless it serves as a conduit for matter. A pipe is therefore a tubular conduit for matter, but one which enhances the gentleman's appearance at parties.

◆ The Society of The Pipe is not akin to the Society of The Spectacles, for while each improves the wearer's gravitas and poise, only the pipe brings both the outside inward and the inside outward.

But life only begins with infancy; after that it often continues for a very long time. Nobody can remain a baby forever, no matter how much certain members of the Žižek family may endeavour to prove otherwise. Eventually one's head ceases to be comically disproportionate to one's gut, and it is time to learn to speak and begin issuing verbal observations about whether crossbeam temperature in Building Seven logically implies a Mossad-led controlled demolition.

At some point, then, one will require schooling. One cannot gabble and mewl until one is sixty-six years of age,[99] and so a program of learning or teaching must be devised. We do not take a position here on whether schooling is necessary; it has been convincingly suggested by some that children would be better off being given pornography instead of schoolbooks, but this seems to us an open question. Rather, for brevity's sake we take the existence of schools as a given, and look at how those schools that *do* exist may be repaired within the bounds of practical possibility.

The American schoolhouse has consistently had a problem, namely that the child's experience within it consists largely of being handed a map and told where Rhodesia is. *But Rhodesia no longer exists*, as anybody who has used the Internet in the last few years has quickly figured out. The education system, then, cannot be said to be functioning to its utmost.

Here, then, are a few home remedies and experimental ointments for the improvement of the American Education System:

♦ Each American schoolchild must be required to memorize at least ten aphorisms from one of the Approved Aphorism Anthologies[100] before being released from the breakfast table for the day.

♦ We support a federal statute requiring paternity testing for teachers, and a link between pay and test results.

♦ Birthdays should be neither recognized nor celebrated in the classroom, as those of us who were born in the summer months always had a miserable time. The collective grouping of summer birthdays together for one party in the last month does NOT remedy the situation.

♦ All surfaces to be used for presentational writing should be black in colour; this is the only colour traditionally held to simultaneously catalyze and symbolize knowledge-acquisition,

---

99 The recently aforementioned exception applies once more.

100 The three currently-approved anthologies are the present book, the *American Anthology of Alliterative Anecdote & Aphorism* (Nimni & Robinson, eds.), and the pioneering work *Sens-Plastique* by Malcolm de Chazal.

with the lightness of chalk illuminating the colossal dark surface of the vacant pupil's brain. All boards of other colours, be they green, white, or otherwise should be swiftly dismantled and stowed in the futility closet.

* Meditation and reflection are inhibited by sensory perception. If all five senses were to be removed from the child, they would finally receive the requisite impetus to follow Doc Soc's command of Know Thyself.

* The United States Government shall hereby undertake an Institutional Redistribution Program, by which free high schools shall be given away to all those wishing to obtain one.

It is our firm desire for these homespun pedagogical fixings to bring satisfaction to the joyless.

EDUCATION, PART II:
ON THE MADNESS OF LEARNING TO BEGIN WITH

"Teach a man to fish and his dependence on you evaporates, along with your ability to control him."[101]

But we must pause to do the backstep in our discursive hokey pokey, for we have made a Knievelesque logical leap over a very deep syllogistic trench. We have, as promised, assumed that it is well and good that schools must continue to exist. But we have also assumed, without justification or explanation, that in those schools, *learning must continue to take place.* It is perfectly theoretically possible, however, to maintain a school while eliminating all learning. Indeed, institutions from Harvard to the European Graduate School have repeatedly demonstrated this to be eminently feasible.

In fact, too much learning can have devastating consequences. For every Chomsky the academy may expectorate, so too does it produce nine times as many McNamaras or Dawkinses. For every time it solves a difficult cryptogram or cures polaroids, the University also produces another disquisition on themes of nation and self in the

---

101 Aphorism #25.

early love letters of James Joyce. The question must be asked: Does the benefit (vast scientific progress) truly outweigh the dire cost (journals of literary theory)?

To the layperson, the mental chaos produced by excessive book-reading has long been clear. We must give credit to Mrs. Victoria "Posh" Beckham, who confessed in 2005 that "I haven't read a book in my life" because "I don't have the time."[102] Since Mrs. Beckham seems to be one of the more peaceable and well-balanced souls we have read about in our time upon that Great Speck formally christened Earth, booklessness must indeed be worthy of serious consideration.

Do books corrode rather than enhance the thoughts of humankind? We are forced to sadly conclude that the answer may be in the affirmative, and that our libraries, like our paintings before,[103] must be thrust upon the pyre.

EDUCATION, PART III:
INTRODUCING CHILDREN TO THE LASSO

"The young people are the most active and vital force in society. They are the most eager to learn and the least conservative in their thinking... the young people should learn from the old and other adults, and should strive as much as possible to engage in all sorts of useful [lasso] activities with their agreement."

-        Chairman Mao Tse-Tung, Introductory note to "A Youth Shock Brigade of the No. 9 Agricultural Producers' Co-operative in Hsinping Township, Chungshan County" (1955), The Socialist Upsurge in China's Countryside, Chinese ed., Vol. III.

But if we burn all of our books, what shall our children pretend to have read instead? What will occupy them and keep them from posing us difficult questions about the dubious justifications for wealth and

102 See a copy of *The Guardian* newspaper from August 16th, 2005 for more details on Beckham's pedagogical philosophy. Note cautiously, however, that this is the same justification posited by or within Sister Ray, and we know precisely what She got up to with all this precious time-time.

103 See Condemning the Arts to the Fire," in Part II, p. 63.

war? How will they be made hardy and brash?

We believe we have found our answer in one simple object, an object which both imbues its bearer with the coarsened prairie spirit and distracts her for long periods of time that can be used by her professor-father for the completion of his multi-volume refutation of Marx's deracialized theory of the commodity.

Enter the lasso. The lasso has reached a celebrated (though dormant) place in the National Mythology. It has given everyone from Roy Rogers to Barack Horses Obama their manhoods, years of childhood practice paying off in the delayed gratification brought by a rich public life. Yet lasso education among the young is at an all-time low, and many of our little golden angels cannot tell the difference between a Honda Knot and a Hangman's Knot by the time they enter grade school.

The disgraces of the present state of affairs must be taken behind the barn and shot. Compulsory lasso lessons are the route to brighter, more pleasing children. No child with a lasso has ever *burned down a factory* or *corrupted a nun*. Each has been polite and tidy, with an aesthetically-satisfactory smattering of freckles and a devotion to God and Family.

Yes, ladies and gentlemen, if you wanted our prescription for the young, you have it in the lasso.

Effects of Lack of Discipline on the Child's Conception of the Rhombus

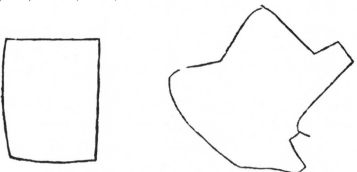

FIG. 122 — RECTANGLE AND RHOMBUS. DRAWN FROM COPIES BY A FIVE-YEAR-OLD GIRL. 1/1.

EDUCATION, PART IV:
FACTS ABOUT THE GALÁPAGOS

But *"youth" is a pseudonym for a far more sinister age.*[104] The real problem is not that children lack a ranch hand's education, but that they are unaware of even basic factoids about the geography of the planet on which they are forced to live. We can eliminate books and introduce lassos, but this will not change the fact that most of us have little idea where we are to begin with.

To this end, let us try an experiment. Here is a short examination on the subject of the Galápagos islands. Please complete it and await further instruction.

> 1. Which nation owns the Galápagos Islands?
> 2. What is the capital of the Galápagos?
> 3. Approximately how many human persons inhabit the Galápagos?
> 4. Do you know of any creature native to the Galápagos other than finches and enormous tortoises?

We expect you received twenty-five percent or less on this examination. This is a failing grade. The answers, incidentally, are Ecuador, Puerto Baquerizo Moreno, twenty-three thousand, and the Waved Albatross, respectively.

This brief bit of inquisitory humiliation is intended to illustrate the problems that come with de-systematization and the elimination of knowledge hierarchies. We will bet money that you knew about the damned tortoises, or perhaps the blue-footed boobies. But you didn't even know what country the islands were in. Or if you did, you at the very least lacked an albatross.

As far as the American child-brain goes, then, where there should be *knowledge* there is only *festering.*[105] If we may deploy an illuminating phrase that stands as the only useful Rumsfeldian contribution to civilized thought: to itself the child is an unknown unknown; it is unacquainted even with the extent of its own ignorance.

---

104 Aphorism #26.

105 "The human brain is like an apple; when dried it becomes vastly more spongy." Aphorism #27.

But how to solve the Galápagos problem? Our personal recommendation is the replacement of learning with facts. Children spend far too much time thinking about why a thing ought to be done and far too little about what it is in the first place. We have de-emphasized the foundations of knowledge, such as the specific date on which the Treaty of Ghent came into effect, in favor of a soupy, feminine insistence on understanding *whether the Treaty of Ghent was a good idea to begin with.*[106]

Thus: do not ask your children if they would like to visit the Galápagos; instead, ask them to meticulously list each bird species present on the island at each point in its history, and punish incorrect answers by assigning additional hours of lasso practice.

EDUCATION, PART V:
IMPROVING THE UNIVERSITY
WITHOUT EUTHANIZATION SILOS

At some point, however, children must stop dabbling in facts and begin going to college. But our colleges have long since ceased to become the rope-trick schools that our frontiersman forefathers foresaw them as being. As optimists, we believe this situation can change. And thus it is necessary to propose reforms.

The last time we made sure, we were not William J. Bennett, Secretary of Education for the Reagan Administration. But we do have various views on how teenagers should be indoctrinated in the social truths. Most of these involve simply shrinking our proposals for other corners of society to fit inside the university, but there are certain aspects of a college education that are different from the experience of, say, having a job.

Yet so many of today's proposals for edifying campus life seem to revolve around extermination silos. March the students in, gently euthanize them (it is said), and the universities will be instantly peaceable and free at last from interminable Student Government quibble-meetings over the propriety/impropriety of the allocation of all campus "activity" funds toward the Utopian Torts seminar's proposal

---

106 We will not here attempt to list each and every fact that a child ought to be taught during their entire upbringing. Life is too brief and history too unremarkable for us to sail the sea of facts in a book-shaped ship. *"Paper boats cannot weather very big floods,"* after all. (Aphorism #28.)

for the construction of an enormous papier-mâché dove in the middle of the Classics Quadrangle.

But though, yes, the silo carries temptation, we feel as if there must be another way. Mass execution has always been an extreme measure, necessary only when called for. If other means to the same end can be found, they ought ordinarily to be taken. To this end, let us tentatively tender a dictate or two, that may help us revivify the university without planting a whole new corpse-forest.

Consider the place itself. Like the beach, the University is in many ways a tiny socialism. Centralized meal plans, complimentary transit and medicine, a life of leisure rather than employment; the rolling campus is a gentle collectivized paradise. There is a reason your two authors have spent the better part[107] of their lives in a university setting. For *isn't it nice* for all people to simply pool their (parents') resources into a vast central fund, and then receive an equal distribution of services, without having to worry at every turn about the earning and spending of particular sums?

Improving the university would thus seem almost impossible. If the solution to everything is socialism, what does one do to solve a socialism other than add more socialism? The real conclusion, then, would seem to be that we ought *not* to be altering the university, but expanding it so that *all other spheres of life were much more like a university than they currently are.* Communal living, shared assets, mass loafing, and complimentary shuttle buses are to be the *new normal* from the airport to the bistro.

Yet several reforms would nevertheless benefit the university. Professors ought to be banned from writing articles, or should at the very least be strongly discouraged from doing so. Practical education must be given central priority, not only the lasso-based arts, but also the care and taming of pachyderms, fundamentals of boatbuilding, and experimental embroidery. Instead of functioning as intellectual penitentiaries, warehousing the young cerebellum for two to four years so that it cannot make a ruckus, the University must craft the immature human into a rounded-off General rather than a specialized Major. The being leaving its walls must be able to tailor a suit, design a water park, pilot a microlight, plan an invasion, defend an indefen-

107   all

sible proposition, woo a barista, locate a tumor, tastefully augment a bouquet, criticize a PowerPoint, toast a retiree, roast a toastmaster, banter with a court reporter, burglarize a film studio, untie a hostage, impersonate a monarch, photograph a pageant, remove a leech, skilfully massage a loved one, file a journalistic dispatch, conduct amateur torso surgery, infuriate a priest, tickle a bellboy, argue with a shipping company, select a tasteful gift, clean a bloodstain, compose a sonata, narrate a funeral, fabricate a telegram, and pen a book of utopian suggestions, among a thousand other abilities.

Other university edifications are due. Gothic architecture is the only form conducive to knowledge acquisition; it must be enforced by law. Professors will cease to insist that students "learn from one another"; a fellow eighteen-year-old has no wisdom of value to impart. On-campus sculptures will cease to be either abstract or monumental; instead they will be both representational and bizarre. All doings will be democratic; matters of university administration will be voted on by both students and groundskeepers. The elderly will be encouraged to enroll, so that students are exposed to more than one another's idiocies. Moving walkways and tunnels will proliferate, as will secret passages and candelabras. Students will take part in semesterly murder-mysteries hosted by the Dean. All students will work, and all staff will learn. There will be little distinction between a student, a professor, and a janitorial officer, except that all will recognize that a professor has spent more time studying the subject and is perhaps to be listened to on matters pertaining to it, and a janitorial officer has spent more time scraping calcified undergraduate feculence and is perhaps to be listened to on matters pertaining to it.

In this ways can an egalitarian yet insistent education be commenced in our higher institutions.

EDUCATION, PART VI:
THE PTA, BUT WITH GUNS

But in addressing education, we have thus far neglected to formulate a corresponding theory of parenting, leaving our argument excessively buttressed but lacking in joist. We pause, then, for some final remarks on how armed parents might contribute to the social good of the young.

★   ★   ★   ★

How many times have we seen that cloying bumpersticker couplet: *"People don't kill guns, people kill people"*? And has it made any more sense upon the thousandth reading than it did upon the ninth? It has not.

We have always believed, as Mao did, that parenting flows from the barrel of a gun. But the gun itself can be literal, floral, or metaphorical. The *essential* point is not the caliber used, but the calibre. The point is that an *undisciplined* education is not one at all, and the parent that praises its infants' fecal smearings as creative masterpieces is disserving both the child and moral truth. It means little to join the PTA if one is not going to be willing to enforce one's decisions…

This is not to say that we are not modest. Throughout this six-part digression we have taken great pains not to present ourselves as experts in the field of education.[108] But we do present ourselves as firm disciplinarians, who can tell a child's social worth from his or her appearance. Our wisdom is the wisdom of the Ancients, handed down from sources sweeter and more majestic than any of the tired adventure stories currently found in boys' magazines. If any of our suggestions are to be taken up, they must be taken up with a willingness to *see them through to the bitter end.*

And yet moderation is simultaneously called for. There is a middle-way in parenting, one that does not tolerate nonsense but is open to literature. Do not be afraid to experiment without thereby becoming experimental. Tell your boy to join the Brownies. Tell your girl to join the Scouts. Buy them a pocketknife and a copy of *The Republic.* But we beseech you, don't buy them tap-dancing shoes, unless you want holes in both your floorboards and your eardrums.

---

108   Each of us has, it is true, published a children's book or two in his time. Parents will no doubt have forgotten Mr. Nimni's *The Black Flamingo: A Book About Fitting In* and Mr. Robinson's classic *Swiss Family Nimni*. But writing a successful children's book requires almost no educational knowledge at all.

What to Do With the Elderly

We have explained the child-rearing process at length. But occasional-ly children survive to adulthood; they do not stay non-elderly forever. How do we plan our society to to deal with the "other" phases of life?

Adulthood itself is a simple matter; the majority of the sugges-tions in this book are intended to apply solely to forty-year-olds, and so there is no need here to discuss the Mature Human. But what of the elderly? The elderly are being widely misused. We have failed to think strategically about how their labour can be optimally exploit-ed, and have instead simply stored them in mass retirement silos to moulden and recline. It is telling that "Grandma has exhausted her useful purposes" was the number-one Facebook status update of both 2008 and 2009.

But the elderly are mostly left idle, like mossy old tractors in a faraway field. They are given procedural dramas and pastel capris, and left to their own devices. They are shuffled out of our cities, mer-cilessly segregated, and ignored until death. When they attempt to speak, we pretend that we cannot hear them or that their words are garbled. When they tell us with sad eyes that they think paying for a decade of graduate school room and board should at least earn them the right to be sent a Christmas card and visited more than once per year, we inform the nurses that they appear to have been given insuf-ficient quantities of intravenous sedation.

Yet leaving aside all ethical questions, this is a crisis of efficien-cy. For the elderly could be put to work handing out chocolates, or punching train tickets. They could pen and recite fables in order to punish wayward children. If we could scientifically reduce their boniness, they might serve as purveyors of complimentary hugs to the lonely. In short, all manner of social purposes might be served through considering the elderly as a *means* rather than an *end*.

The country has the resources to mobilize the elderly. It man-ages to get them dressed each morning and put them in front of the television, it manages to gather them up and store them by the millions in Floridian nursing-warehouses. And yet somehow we have not managed to turn the elderly into successful soldiers and teachers, despite the fact that they are most cognizant of what life is and have the longest-winded of all anecdotes. The oversight must be corrected.

What Dating Should look Like

But in order to have satisfactory elderly persons to begin with, we must ensure the quality of the process by which the sexual reproduction of young people occurs, since each elderly person emerges originally from a womb. By dealing with errors in the dating process, we might forestall many of the problems that occur among the old people who were born to the dating couple.

When two humans wish to copulate, and produce love, the resulting process is more or less random. I date only those people whom I have heard of, or who live in my town. I date without any particular strategic maximization of any end-goal in mind. The result, as one might expect, is disastrous. Coital thoughtlessness leads to both suboptimal production of pleasure-units and poorly-designed offspring.

The situation has improved in recent years thanks to the internet's automation of the dating process. At last mathematics is being given its due place in the assignment of mates. I am not told what a potential partner is like to play backgammon with, or how a conversation about Foucault over *mojitos* will go; instead I am given only a percentage, a pure distillation of quantified compatibility.

But a problem remains that no algorithm, however complex or brutal, can exterminate: voluntariness. Dating remains voluntary, and until it ceases to be so, society will be failing to ensure ideal reproductive partner-assignment. For what stops me from accepting a 68% match over a 92? Only my dignity and the fear of the inevitable hideous offspring. But the thrills of intercourse occasionally lead to decisions that fail to take into account their full range of multi-decade outcomes. Thus, granting individuals override power, and allowing them to circumvent the formula, while it may serve the abstract value of "liberty," is easily outweighed by the values found in mandatory algorithmic deference.

The problem with arranged marriages is that they were poorly arranged, with imperfect information. They were before their time, an act of hubris on the part of humans who thought they could decide for other humans what only mathematics was capable of knowing. But arrangement itself remains desirable, and as we slowly approach the technical capacity to fulfill the promise of early rudimentary arrangement regimes, we should not allow irrational humanistic concerns to circumvent sexual efficiency. No one is capable of choosing her mate well, or at the very least, those who are capable of choosing their mate are outnumbered by those who will choose their mates less well than an intelligent algorithm; thus maximizing human happiness necessitates eliminating the choice component in dating and turning to a system of automated assignment. Anything else needlessly sacrifices units of pleasure, and is therefore not only bereft of reason but unambiguously cruel.

★　★　★　★

Yet our proposed dating platform fails to address the foremost reproductive dysfunction of contemporary erotic life, namely *the reduction of information loss during sexual intercourse*. Intercourse itself is often categorized as a "closed system," through which particles of pleasure and information travel freely from unit to unit. But in practice, this is untrue. Distractions such as televisions, books, and text messages necessarily interfere in the act, initiating a chain of events whose inevitable conclusion is the shameless elopment of one's disloyal spouse with a gormless local dental professional.[109]

The problem of the age is the problem of Information Loss. Romance has always emerged in various ways, from convincing a nun of the necessity of atheistic hedonism to placing wry personal ads in the *New York Review of Books*.[110] But internet tubelets have dealt a death blow to dymaxion sexual processing, and the change is reflected in our own bodies. *At least one partner in any contemporary carnal encounter is inevitably distracted*, and if one of the two is an ethnomusicologist and the other a sociologist, it is *never the sociologist* who fails to perform.

Far from the ideal insemination tactics perfected by the so-called "Old Right" or the love gods of the Hindu myths, we now exchange genetic material listlessly, in a way that leaves half-formed children forever wondering about their true identity. Without efficient reproductive information-processing, can a child truly be said to have a "nature"? We answer in the negative. Open data channels hemorrhage

---

109　whose own teeth, ironically enough, are conspicuously aesthetically subpar in spite of a (suspicious) regional reputation for professional excellence

110　"Fusty A-student seeks tweedy Cold War Liberal for quiet disagreements over Lionel Trilling"

values, thus retention of values requires sealed organs. No sexual act can be engaged in *scientifically* without complete bodily enclosure in some sort of information-retention "bag" or "bubble." Through this process of total data immersion, our fornicatory acts may gain both precision and wisdom, and ennoble the human spirit. If sex were conducted in sealed containers, the number of marriages salvaged would be incalculable.

FORNICATION AND THE NEW AMERICAN CENTURY

It is difficult, however, to announce a position on dating without correspondingly addressing the question of whether and how people should fornicate.

Of course, evasion is the usual tactic; most philosophers mention fornication little in their works. From the way it is presented in scholarly texts, one might assume that fornication died out with the '60s. Not so. In fact, it is as rampant today as it was in those heady and chaotic times, especially among certain more lithe members of the so-called teenage class. We have even watched our own students so consumed in the act of giving one another the Eye of Desire that they neglect to take notes during crucial segments of that week's Utopian Torts lecture.

But must they be stopped? Never. Bodies are designed to slide within and without one another, to create shapes and twirls, and to vigorously and splendidly join and separate. We encourage the growth of this trend, and only regret its non-existence during our own monotonous youthful days. "Go forth, young humans," a Professor might reasonably say to his flock, "and do as you please to one another!"

This truth has not been given equal quarter from every political faction. When William Kristol and the founding members of the Project for the New American Century declared in 1997 that "We are in danger of squandering the opportunity and failing the challenge... We seem to have forgotten the essential elements of the Reagan Administration's success," they referred through somewhat obfuscatory language to the fornication of the underclass. But not all of us have the Ivy League diploma or abundant social capital necessary for the maintenance of a Great Books lifestyle. Not all of us can love like a neoconservative. Not all of us can keep our libidinal urges in check or maintain perfect fidelity to the national virtues. For us, there must be an alternative, and whether the PNAC den-mothers care for it or not, this alternative is a fornicatory one.

We do not, of course, support returning to the unchecked and relentless shaggery of certain former decades. But we do believe that Kristol and his ilk should be tagged as "out of touch" with the people's sexual necessities, and lambasted thoroughly in multiple media. This work can be counted as one such lambasting. "Shame," we say. "Shame on all those who take no pleasure in watching their undergraduates make love."

Men & Women

But so many speculations on intercourse and proper dating procedure are premised on the existence of men and women, a premise we oppose. For we have always felt genders to be an untold silliness; why as a male professor am I expected to wear a necktie instead of a leopardskin brassiere? The question is impossible to answer satisfactorily.

The capitalist answer to the gender question has always been: "If you don't like your gender, change it." But while we, like our capitalist comrades, have always maintained a basic skepticism toward Stalinist thought and unnecessary prescriptions or prescriptives, we find ourselves unable to fully climb aboard the sloop of liberty in this instance. Our own formula varies slightly: "If you don't like your gender, throw genders into the sea and refuse to speak of them again."

We believe in the masculine and feminine. We have always enjoyed both stripteasing in fishnets and roleplaying lumberjacks. But we cannot sign onto the wild essentialized notions of "male" and "female" which have so calcified themselves in political discourse. Everyone enjoys breasts, it's true, but to feel as if one needs breasts in order to be feminine strikes us as a mathematical mistake.

Thus, if asked for our position on men and women, it is thus: there should be no men and women at all, only *us*.

If Birds Had Facebooks...

In is unfair, however, to speak so anthropocentrically of men and women, when there are in fact genders to be found among many animals, and there are, indeed, many different types of animal to begin with.

This section of the book, then, will be dedicated to tidying up some of the uncertainties we have previously left in our work about birds. When reviews of the first edition came in, the number of those who were puzzled by our failure to mention birds almost exceeded the number of those who were not. Where matters avian concern themselves, we have been accused before of serial vagueness.

But if our views on macaws seem somewhat evasive,[111] akin to the cannibal's coy vegetarianism, our views on birds as a *sum total* are far from the opposite. In fact, some of our strongest positions on the matter are to be found in the very hypocrisy of the prior sentence. For we do *not* believe birds to be quantifiable as a mere "sum total" in the way one might speak of a barrel of tar or five tons of flax. Each bird has a distinct and rich personality (birdsonality) and to treat he/she as but a fungible cog in an enormous worldwide bird-machine is despicably dehumanizing (debirdably debirdenizing).

There is one solution that would instantly resolve each and every problem facing contemporary birds. If each bird were given a fully-functioning Facebook account, it is safe to assume that social networking itself would become overburdened and inutile. If for every time one tried to locate a long-lost grandparent, one was instead faced with the prospect of having to scroll from Pigeon #753434 to Emu #4375223, we might accomplish multiple objectives. First, the inconvenience would be so great that human beings would rapidly abandon the internet, forcing them finally to return from the simulated reality to the real one (though alas, not from that subsequent simulated one to the actual real one.) Second, further assumptions (to be made despite our full awareness of the general caution against such constructions due to their tendency to render "asses" of both profferer and recipient) lead us to believe that Bird Facebooks would

---

111 See Nimni & Robinson, "The Indefinite Macaw: Blurring the Bird, Burdening the Blur," *Audubon Society General Newsletter*, Vol. 120, Iss. 3. See also Various Authors, "Replies to Nimni & Robinson's Indefinite Macaw," *Audubon Society General Newsletter*, Vol. 120, Iss. 3.

restore to our wingèd sisters the vast majority of the dignity stolen from them during the Great Migration. Birds have too long been the subject of humor and *New Yorker* cartoons; it is time to afford them full coequal status as people instead of punchlines.[112]

KITTENOUS UBIQUITY

The bird is not the only animal. An anecdote from the reactionary papers:

> *"I'd like to say a few words about what lies immediately ahead, and what this may bring to the Kurdish populations of the Middle East," Chomsky said. "It's clear that the government of the United States, with Britain trailing along, is desperately seeking to go to war with Iraq, although the disparity of force is so vast that the term 'war' is hardly appropriate." As he spoke, a tiny orange kitten appeared and wandered out in front of the stage. It spotted the huge audience and froze, terrified. Several of the photographers snapped pictures of it. It ran back and forth frantically and then hid behind a curtain. "Like most states in the world," Chomsky continued, unaware of the kitten, "Iraq is an artificial creation--it was patched together by the rulers of the world eighty years ago in order to satisfy two conditions: first, that Britain, not Turkey, would gain control of the huge oil reserves of the north, and, secondly, that the British dependency of Iraq would have no access to the sea and therefore would remain a dependency. When the United States took over global management from Britain sixty years ago, it kept the same arrangements in place." Another kitten ran out onto the stage, followed by the first one. The two curled up together and fell asleep.[113]*

If even enlivened is a Chomsky lecture through the supplementation of kittenry, surely a timeless lesson somehow drawn could be? Yet,

---

112 Even the introduction of such "Facebirds" is unlikely to fully harmonize human-animalian relationships. Until a formalized global peace accord, with similar terms to those reached decades ago with the fungi, is drawn up and agreed to, our mass exclusion of birds from political decision-making is likely to persist.

113 *The New Yorker*, sometime in the last 20 years.

besides Israel, how many Western nations have included the provision of wandering street-kittens as a core part of their social agendas?

The common objection will inevitably arise: "But, Mr. Robinson, how can you suggest the addition of new kittens when we all know the regrettable history of Horse Zionism?" The obvious rejoinder is that we do *not* all know the history of Horse Zionism; in fact it has disappeared from many lesson plans entirely (another tragic unintended consequence of 9/11). Children are today being raised with no knowledge of what can go wrong when non-human animals are given full decisionmaking power over the military apparatus of a state.

We need not go into that background here, for the calibre of reader willing to spend twelve dollars on a Nimni-Robinson album will have no need for political-historical babble about horses. Interested dilettantes may consult the internal diplomatic record[114] at their leisure, if leisure indeed can such an activity be called.

Kittens differ in important aspects from the horse, and we boldly propose that a kitten-heavy state would depart considerably from the authoritarian butchery of experiments in equine rule. Readers may consult other chapters of this work for insights into why an exceptionalist theory of horse politics might be justified.

In fact, it remains true that little harm has historically resulted from increases in the number of kittens in a place. Certainly, there is a theoretical level at which the presence of kittens becomes intolerable. Five in the lap is endearing, but fifty on the face? Debatable. If the earth's entire mineral resources were extracted and directed towards mass kitten production, the resulting vast mewing swarm would favor no party's interests.

An excess of the feline is thinkable, then. But the reader should not be distracted by extremes. Too many kittens create an invariably slippery slope. It is not necessary to think of a thousand kittens when one may think of several instead. The reforms we propose are moderate; if we are about one thing, it is subtlety and restraint. Add some kittens to your diet, but do not base your whole digestive curriculum around them.

114   in the original Hebrew

THE BUICK AND THE RHUBARB PATCH

[The following chapter was recommended for deletion by our book's first editor (since resigned), who felt that it disrupted the Part's narrative and offered a questionable contribution to the overall theory. While we took this suggestion under advisement, we came to the ultimate conclusion that it was premised in a misunderstanding of the nature of both narrative and theory.]

But when we are dealing with a prescription for adequate numbers of kittens, we are ultimately dealing with the question of *how to achieve a well-balanced society.* That necessitates discussion of non-kitten-based subjects just as much as it necessitates discussing kitten-based ones.

To this end, let us pause to examine the case of Phoebe. Phoebe is an ordinary human. Phoebe drives a Buick. Phoebe is a good-hearted woman and a sincere patriot. Every month, she pays her dues to the National Bricklayers' Union on time, although she is not herself a bricklayer.[115] She buys American, because she remembers Pearl Harbor.[116] She hates children, but she loves church. In short, she's you and me.

But, as you will have seen, there is a problem with Phoebe's lifestyle. It has no *dessert.* These United States were founded upon the ideal of the main course (the joyous Thanksgiving myth, the inspiring rise of the hamburger stand and hot dog cart, etc.) The national consciousness shuns the sweet. When George Washington cut down his cherry tree, it was perceived as an injustice, for to obtain as sweet a fruit as the cherry is a violation of the American gastro-ethic. Americans are hardy industrious Protestants; here confectionary is considered frivolous and European. Take, too, Thoreau's huckleberries! Thoreau loved those huckleberries, and was jailed for it! And let us never forget that little Johnny Appleseed was shot for daring to spread apples across the land.

The point here is that America has, whether justly or unjustly, historically favored the savory over the sweet. But we feel that a compromise can be forged.

Enter the rhubarb. The long-neglected rhubarb creates pies which

---

115 Note Aphorism #29: *"There are far more people laying bricks than there are bricklayers."*

116 Misspelled "Pear Harbor" in an earlier edition of this work. Apologies to the families of victims, who may have felt trivialized. Fruition was never our intention.

resemble a traditional dessert, yet are bitter and nasty and must be subjected to endless sprinklings of sugar before reaching a state of even mild palatability. It has all the advantages of a traditional American pie after-course, but with the sensory qualities of a traditional entree.

We feel that the rhubarb could provide significant benefits to the people of this country, and support its mandatory adoption through means of legislation.

What Shape a Bagel Should Be

Yet let us abstract ourselves somewhat. When we propose the spreading of rhubarb, what we are really discussing are *essences*. The rhubarb's mixed messages of sour and sweet come from *the nature of the rhubarb*. But would a rhubarb that was solely sweet therefore not be a rhubarb? Where does, say, the rhubarb end and the sugar-radish begin? What does it mean for a thing to be itself instead of some other thing? For example, if one's mother had never once woven, was she never a weaver? If the Boston Independent Shakespeare Festival showcases plays that were entirely written by two local professors instead of Shakespeare, is the resulting legal injunction and asset seizure necessarily warranted?

The discussion of essences therefore invites the title question: What shape should a bagel be?

Our answer: **round**.

This contention, that roundness is a necessary condition of bagelness, is more radical than it may seem. It puts us at odds with some more "progressive" professor, who have denounced the concept of fixed essences as little more than the first paving-stone on the road to authoritarian rule. If avant-garde bakeries wish to dabble in the trapezoidal and square bagel (squagel), who are we philosophers to tell them they must forgo the right to use the storied name? How can we prescribe what is made by doing?

But we have never let popular disagreement keeps us from speaking truth. New experimental bagel prototypes have had their day, and the time has come to return to first principles. We must interrogate the question of purpose: if a woman cannot see through her bagel, what is it for? The centre-point of the bagel is the opening through which all life flows, and those bagels that have closed-up or misshapen holes are to be both discarded and condemned.

The central tragedy of the modern luncheonette has been the proliferation of lax bagel standards. We have proposed corrective legislation for this crisis, but have as of this writing been greeted with little more than polite buffings and rebuffings from those we find hanging around the Halls of Power.

★   ★   ★   ★

But there is another principle underskirting our position on both essences and bagelry. It derives from a larger macronumerical maxim, namely *The Pleasingness of a Nice, Well-Rounded Number.* After all, we should remember two relevant facts:

1. Nothing tickles the eyeline or fingertips more than a number with a few 0's in it.
2. The single most compelling argument against the admission of Puerto Rico as a United State is the irreparable damage it would do to the "Fifty Nifty" song.

The influence of Essential Form on the mind is by now too well-documented to require documentation. From the results of the psychologists' investigations, we now know just how crucial presentation is to reception. Hence, the book before you is presented in a rectangular format, in order to convey simultaneously the dual messages that The Authors Are Not Squares but also that We Know Our Boundaries. If we had released a circular book instead, you may think us mad, even if you were generally a sufferer of open-mindedness and Political Liberalism. In spite of the unparalleled delights of the 0 where matters aesthetic and statistical are concerned, it is not a form with universal applicability, and its overuse must be cautioned against. Bagels, numbers, blueprints: each has a perfected form, deviation from which lessens the fulfillment of its earthly task.

Given how certain we appear to be of this, can there be any lingering question that our bagels ought to remain round?

The Hunk of Butter

But few people would trust an ontology that had not been seconded by a folk singer. Witness what Mr. Tambourine Man himself had to say about objects and essence:

> *"To say 'cause of peace' is just like saying 'hunk of butter.' I mean, how can you listen to anybody who wants you to believe he's dedicated to the hunk and not to the butter?"*
>
> - Bob Dylan,
> Interview with Playboy Magazine,
> March 1966.

Though we detest Bob Dylan as much as the next professors, as the original coiners of the classic watchmaker's aphorism "Even a broken clock is right, if one really thinks about it," we must concede that Broken Clock Bob is bang-on about the hunk of butter. A thing must be about that which it is about, and nothing different.

(Of course, March 1966 couldn't have gotten here soon enough; the moment we read the above *Playboy*,[117] we realized that we had already signed several anti-Vietnam petitions strenuously advocating, in no unexplicit terms, "a firm commitment on the part of all political authorities to the realization of the hunk of butter." Certainly, no more egg has ever been on any two faces. But this bit of humiliation aside, Dylan's diagnosis is to be fully embraced.

The hunk of butter is *precisely* what Vietnam was not about. In a time of Brokawvian starry-lipped revisionism, it is more crucial than ever to establish raw, encrusted historical *factitude* as against generalized romantic sweepery. It is forgotten that the much-eulogized "Greatest Generation" beat their wives and had horrendously wide taste in lapels.)

What Bob Dylan says about the Hunk of Butter, then, is simply a restatement of our own position on breakfast foods, albeit made fifty years previous. We would like to think he would have credited our work had he anticipated it.

---

117 We will admit to maintaining an active subscription to the periodical, but insist that such maintenance occurs only for the limited purpose of indulging our purely academic interest in masturbating vigorously to pornographic images.

But instead of dwelling on decades-old academic feuds, let us now change the subject, and discuss the matter of dignity: who has it, who doesn't, and which high-profile Slovenian philosophers may have sacrificed it through their second-rate scholarship.

FINISHING YOUR DRINK AS THEY DRAG YOU AWAY

How does the human animal relieve his despair? Sex may be the only *medically*-recognized apparatus, but what of herbal teas and the other "alternative" cancers? Judging by the number of people who cite "chai" as a pastime, there must be some consensus amid the madness, no?

No. There is one ointment alone for despair, and it is dignity. "But," says the onlooker, "What is the value of dignity when one is essentially a chimneysweep?" Nevermind your damn chimneysweeping! The chimney is only a chimney in accordance with one's perceptions. Reimagine the chimney as a birth canal, the soot that cakes your lungs daily as the glimmering placenta of the new society in its infancy.

Dignity it is, and dignity it has always been. To die on one's knees rather than live on one's forceps, *this* has been the sole demand of the tortured workingman from Athens to now. Martin Luther King knew it, and so did Big Jesus himself. Dignity in the face of reason shows not only the presence of virtue, but the absence of serious vice.[118]

Here we find the dissident: persemacuted and ostrich-sized. The dissident can be assured of one thing alone: that she will be crushed. With this kind of certainty abounding, she faces only one serious decision: what faces will I make as they pull me from my chair and throw me on the heap?

We posit the following: the act of rebellion is consummated not when one is dragged away, but when one exhibits the composure and

---

118 We might give further useful examples of dignity and humility by discussing our own. Our well-known draft email "On the Seeking of Vengeance Against Those By Whom One Has Been Wronged" is compellingly illustrative. In it, we meticulously and profanely replied to all of those among our peers who slighted us by referring derisively to our work on this book ("Penning the Bloops, are we?" was the sarcastic inquiry dealt endlessly to us in hallways over six maddening years) and suggesting we had deprioritized our teaching commitments despite having that very semester prepared a syllabus running over two dozen pages. Our restraint from sending this masterful philippic to the faculty listserv was, we feel, a very paradigmatic pinnacle of Dignity in Action.

arrondissement necessary to *finish one's drink* as they drag one away. In those last sips, taken with calculated obliviousness to and disdain for circumstance, one has truly smeared tomato on one's accuser.

It's not a wondercure, certainly, but the question "What else can you all do about it, anyway?" has never been met with a *cosmically* good answer. We are do-makers, and so we make do. Grist for the isthmus, as they might say.[119]

Not every answer will lead to satisfaction, just as not every conniving sexual encounter will lead to a tenure offer. But we only aim to say things that are *mostly* true, and it is *mostly* true that Finishing Your Drink is the best thing you can probably do as they Drag You Away.

THE FISHERMAN-TAXIDERMIST

Dignity has a wellspring, however. One is dignified in proportion with one's fulfillment or non-fulfillment of one's personal inner selfhood. If I want to build lighthouses, but I am stuck flying planes, I am alienated and undignified. If I want to operate a lighthouse, but my family insists I go to graduate school in sociology, I am imprisoned by my material conditions.

Mr. Marx may have put this dilemma best:

> *"[In] communist society, where nobody has one exclusive sphere of activity but each can become accomplished in any branch he wishes, society regulates the general production and thus makes it possible for me to do one thing today and another tomorrow, to fish in the morning, taxidermy in the afternoon, and taxidermy again after dinner, as I please, without ever becoming either fisherman or taxidermist."* - Marx, <u>The German Ideology.</u>

---

119 See below:

grist

isthmus

For more, see our topical monograph *Hegel's Isthmus: A New Topography of Philosophy*, 1000 hand-stapled copies of which sit in an appropriated milk-crate in unused corner of the Faculty Lounge.

The ideal, then, is that instead of having professions all things become pastimes, that instead of careers all things are callings. The unification of professions, so that humans cease to feel limited and squeezed to death by their specialized rôles, is a desirable next phase of social life. We will do as we please in the morning, then do as we please again after dinner.

But where Marx fell short was in replacing occupational division with temporal division. What if one were a fisherman in both the morning *and* afternoon, but also simultaneously a taxidermist? Why are we being forced into a choice between what to do now and what to do later? This, surely, is food for thought.

Our own concept of the fisherman-taxidermist resolves this problem. He is always fishing, always taxidermying, never doing before one meal what he does not do after another. The fisherman-taxidermist is an enriched form of each of his two occupations, his forearms more firmly plunged into the meat of his practice. Like the philosopher-king, the true fisherman-taxidermist is not afraid to get his hands dirty.

There are criticisms to be made. Certainly, we think carefully of what Lewis Mumford meant when he said that "[t]ailors and tinkers, almost by definition, could *not* be humanists" (emphasis emphasized in original). The aimless pursuit of little pastimes does not build a United Nations. But is not to tinker human? To tail divine? How many little men with measuring-tapes have fitted us for how many waistcoats, *and when they did so were they not upholding their humanity with greater grace and purpose than all of the soldiers in all of the trench-craters in all of the War Memorials?*

One need not dabble in entrails to have moral worth, of course. There are honorable professions that do not involve the removal of an animal's insides. But we are reminded of the old phrase "You don't have to be a taxidermist to work here, but it sure does help."

There is as much spice in variety as there are varieties of spices. But it is not always advisable to salt one's wounds, and there are more things that one can be than a mere fisherman or taxidermist. Still, it is wise to bear in mind the core moral, that we are ultimately always attempting to round ourselves. Humans should not be poked into tiny holes, but should blossom into enormous all-encompassing adaptabile gelatins. Fish, taxidermy, and be merry.

Dantleyism

But the living of a utopian life in a dystopian neighborhood is not so simple a matter as merely obtaining one's taxidermists' license. There is an *attitudinal* shift necessary, one which prioritizes the performance of meaningful wholesome tasks over the chasing of material spoils.

Yet rôles require rôle models. It is very easy to say "Go and be a different kind of architect," but it will be all but impossible to comply with this new law unless I have been handed a pencil. Who, then, are we to emulate, so as not to have to come up with our characters by ourselves?

We believe an answer may be imminent. Read:

> *Former NBA star Adrian Dantley spent years guarding opponents on the court. Now he's guarding schoolchildren as they cross the street... Dantley, a hall-of-famer and former star for the Utah Jazz and Detroit Pistons, started working as a crossing guard in September. He works an hour a day at Eastern Middle School and New Hampshire Estates Elementary School in Silver Spring, Md. Dantley grew up in the area and says he took the job for the health care benefits and to have something to do. Montgomery County civil service records show he gets paid $14,685.50 a year. Dantley says he doesn't need the money. He says he enjoys giving the young children high fives and encouragement.*

At last, there is hope in Dantleyism! Look here at what we have: proof that willpower can overcome expectation.[120] Need a player of basketball lead a basketball-player's lifestyle? Not if when a man steps off the ball-court he instantly embraces the Dantleyist worldview. For some time we have ourselves struggled to articulate a philosophy of "Confucianism for Basketball Players," and here in A.D. we believe we may at last have found our exponent.[121]

But the Dantley-tale begs a question: must one have played professional basketball in order to be kind to children? Certainly, our own brushes with the toddling class have convinced us to deprioritize

---

[120] *"We are all but pilgrims in search of expectation; insanity our lighthouse."* - Aphorism #30.

[121] Those interested in further study of our attempts at crafting custom-tailored ontologies for sportsmen may wish to download our limited-release e-book, *The Cricketer's Metaphysics: Episteme Beyond the Pitch*, an attempt at inviting the uncultured yeoman sports player into the philosophical enterprise and simplifying complex intellectual inquiries to make them accessible even to the athlete, which we consider firm evidence against charges of our academic elitism.

reproduction as a social activity. Our outputs have consistently tended toward the scholarly rather than the natal.

The problems with children are, as we see them, manifold. First, each course must have its prerequisites, yet children come equipped as curious, empty-headed *tabula rasa*s, lacking in the necessary prejudices and excessively inquisitive of the premises underlying mature thought. Try to explain a finer point of post-Kantian representation theory to a child, and instead of soaking up one's reasoning like an eager cheesecloth, the child is likely to posit inane and unproductive side issues concerning the reasons for studying post-Kantian representation theory in the first place.

Our own suspicions about children were confirmed when we made the mistake, encouraged by the Dean as part of a scheme to indebt younger consumers, of opening up our "Critique of Erotic Reason" course to children under ten. The results were catastrophic. While doctoral students and advanced undergraduates were perfectly happy to believe that erotic reason both existed and could use critiquing, the "littl'uns" (as they were called in enrollment brochures) diverted the entire semester into an exasperated iteration and reiteration of the arguments for the practical necessity of philosophy seminars. Children, then, are problematic.

But Dantley is not thereby undermined. Helping them to cross streets remains an endeavor to be complimented, for though a child may be an irritant, it does not therefore deserve to be ground beneath the wheels of a delivery-truck. Safe streets are sacred streets. As Dantley speaks to his charges in the new currency of the High-Five or its evolved counterpart, the High-Seven, he speaks somehow for us all. Hope in Dantleyism, Dantleyism therefore to hope.

Untying Your Shoes
Without Gouging out Your Eyes

*Paul Schaefer was born near Bonn in 1921. He had a glass
eye, having accidentally gouged out his right eye while trying to
untie a shoelace knot with a fork. He joined the Nazi youth
movement before becoming a Luftwaffe medic stationed in
France during World War II.*

"Paul Schaefer, 89, ex-Nazi preacher jailed for abuse,
dies." <u>The Washington Post</u>, April 27th, 2010.

One must not only change what one does, but one must also change
how good one is at it. In an age that glower-frowns upon true crafts-
manship, the old trades have been worryingly neglected. Once a man
would deliver the seltzer water to the door each week; he would meticulously adjust it to optimal carbonation quotient and place it in
hand-blown glass bottles. The seltzerman, the milkman, the jam fairy,
all have found themselves mashed beneath capitalism's ruthless ten-
derizer. Who these days can stay in the singing telegram business?
What market is there for customized umbrella handles? Personalized
novelty mayoral sashes? Very little, as we know only too well from
a series of humiliating forays into these industries, which not only
embarrassed us in front of the academic colleagues to whom we had
prematurely boasted of our inevitable prosperity, but also managed
to exhaust the entire vast Nimni family shoehorn fortune in a mere
matter of weeks.

It is hard out there, then, for a man who not only wishes to do his
job, but wishes to do it well. Today it is enough simply to trowel the
grout without glazing it. A man renting me a boat can simply hand
me the boat, a typewriter repairman will merely repair my typewriter.
Those extra touches, the *lagniappes* and flourishes that constitute the
charm of true *love*-craft, these each are streamlined and have their
eyelids sanded off.

Artisanry is dead, then. It survives only in name, in the form of
bearded entrepreneurs charging $7 for a gourmet muffin with unex-
pected ingredients. But artisanship is not about discovering the maximum price that can be charged for the minimum size of pastry. It
is about becoming the kind of individual who does all things with

honed and patient technique, who carries the skill of successfully un-tying one's shoes without gouging out one's eyeballs. In view of this, we now wish to take a short amount of space to offer some tips as to how one might untie one's shoe without gouging out one's eye(s).

The true key to not stabbing yourself in the eye with a fork over a particularly nasty shoelace knot is to avoid the appearance of the knot in the first place. You could wear loafers! Or you could tie your shoes rather loosely. They may come undone, and you risk tripping, but at least you will still have plenty of eyeballs! If you *do* get lumbered with a knot of the particularly intractable and *Schaeferesque* variety, the first thing to do is *stay calm*! Panic only alarms the knot, and can cause it to defensively tighten into an even greater state of undisentangleability. Gently coo at the knot to get it on your side. Do not stab it with a fork, not only because of the eye-thing, but because it might hiss at you. If the coos do not do the trick, it is likely that you have a Sinister Knot on your hands. These knots will not be reasoned with. Better burn your shoes and destroy the ashes.

TRUTH IN DENTISTRY

But we mention successful knot-untying mainly to discuss *moral char-acter*, and it is impossible to discuss moral character without discussing dentists...

All dentists are, in some sense, liars. Each conspires to maintain the falsehood that our teeth will not someday fall out and rot away, and that it is therefore worth the energy spent to "take care of them." The dentist is constantly foisting gaseous untruths down our throat, from the idea that he is our friend to the myth that we must floss.

Of course, in some sense the dentist is an easy target. After all, the encyclopedia entry on "Memorable Dentists" is among the more tumbleweed-laden of the Ma-Mn volume. Today's Denture King of Toledo will tomorrow be forgotten even by those who owe their highly-envied toothy grins to the DKoT's artistry.

But easy targets need not not be shot at. We have never subscribed to the mad humanitarian postulate that fish are entitled to clemency simply because they inhabit a barrel. Does the dentist's status as an insignificant nebbish offer mitigating evidence in his favor? It does not. The dentist's betrayals must not go unavenged...

We pause here for a procedural aside. Colleagues reviewing draft copies of this chapter have been somewhat critical of our approach to the subject matter. Most of the comments we received implied that our inquiry was somehow being conducted in bad faith, that it seemed as if personal animus rather than objective necessity was undergirding our position on dentistry. The charge, of course, was offensive; a scholar does not like to have his motives impugned, not least by those with inferior publication records.[122] It is true, of course, that each of us has at least one former spouse who is currently romantically associated with a dentist. It is true, also, that the dentists in question happen to be wily prevaricators who are bad at their jobs.[123] But a true academic does not allow the personal to invade the scholarly. If dentists known to us personally happen to be the kind of untrustworthy tooth-finaglers described in the above, this is only further evidence in support of the theory, and not the other way around.

Our reviewers also unfairly indicated that the posited notions smacked of pettiness and disproportion. Are the dentist's crimes so civilizationally shattering as to merit such thorough treatment? We take this question more seriously; certainly we do not believe the dentist is of any consequence and hesitate to pay him any attention, for we spend so little of our time contemplating his past effects on our matrimonial wellbeing. But we vigorously dispute the idea that disproportion alone invalidates a theory's worth. In fact, some of our bitterest disputes have been over trivial matters. Many of our theories emerge from "personal" issues that some would classify as "insignificant." Such incidents are valuable because they expose the operation of *principle* at the micro-level. They are *object lessons* that allow for the building and testing of larger macro-level concepts. To go from the individual incident to the generalized applicable principle, this is the fundamental empirical process of sound scientific inquiry.

Dentists, therefore, are an unwelcome rancid spume on the cleansing wave of the medical profession. The world would not be worse off for their deaths.

---

122 Ordinarily, reviewers of first drafts are kept anonymous, but for the sake of full disclosure we will indicate that in this case they were Prof. D.D.M. "S." Abraham of Deer Country College in Berkeley, California and Prof. Omri Nimni (no relation) of New York University.

123 One of them, we have on solid second-hand knowledge, graduated in the bottom tenth of his dental school class.

C. Political Arrangements

UNDERMINING THE LABEL INDUSTRY

Among young Leftists of the more senselessly radical variety, the burning down of factories has undergone a recent upsurge in popularity.[124] As stalwart supporters of both the rule of law and the sanctity of industrial production, we reject these infantile tactics, *yet we can appreciate and understand the motivating forces behind them.* It is not, however, the entire picture.

★   ★   ★   ★

The problem of the age is the problem of Labels. An enormous quantity of objects are currently given labels, from social groups to jam jars. An entire industry has developed in support of these endeavours, finding its foremost profiteers in the Consolidated Label Company and the Whitlam Label Company.

Naturally, as theorists concerned with the dynamicism, variability, and unclassifiabilty of human life, we reject these attempts at authoritarian classification. No object can truly be considered "99% Juice" and no person can truly be considered a "hipster" or "beatnik," just as no clear dividing line can be drawn separating the hand from the wrist. Given this fact, we support the efforts currently being under-

---

124 See the following news articles for more information:
   "Two teens charged with factory arson." The India Times. Feb. 20, 2010. Blundell, Kay.
   "Arson fear as gnome factory burns down." The Dominion Post. July 14, 2009. Johnson,
   Mike. "15-year-old charged in Burlington Coat Factory arson." Milwaukee Journal-Sentinel. Feb 12, 2009.

taken by both relativist academics and beflanneled college freshmen to eliminate (or at least significantly weaken) Labels and their counterparts.

In the course of this mission it may be tempting, of course, to fall back on the tactics that have come into vogue among other political demonstrators: factory-based arson. With so many factories churning out so many sheets of so many labels on so many days, who among us could resist the urge to set a fire in the lobby or defile the newly buffed shop floor?

Yet we caution you: these tactics are those of the Scoundrel and Street Menace. Labels must be confronted on their own terms, through Careful Peeling and Obscuration By Magic Marker. In this way, we may avoid the needless imprisonment of countless frustrated and identity issue-stricken teens, but may also find that the next time we purchase a can of soup, we cannot read the fat content.

FINDING THE ORIFICE FROM WHICH POWER FLOWS

But of course, burning down a factory is only one route to the just society. Power blooms in a thousand deathblossoms, and so one thousand hoses are needed to extinguish its tendrils. Finding the source of power means finding the hope for its undoing. As the saying says, when power to flattery bows, power to battery flows. But what flows toward power, or to put it another way, what do they actually mean when they use the word "power"? Can all power be measured on the same scale; i.e. are torque and horsepower and the presidency coterminous?

We think not. Instead, we think power exists in a series of interlocking matrices, each of which entails the other. For some time we have vigorously attempted to convey our theory of power to colleagues, only to be disinvited from future departmental picnics. But is their ignorance excusable? At the time, we told them it was not, but time reconsiders a wound. It is now our position that those who deride our theory of power as "incomprehensible" simply do not comprehend it. This may partially be due to an atypical use of the word "matrix," which we use to denote something quite separate from what it means in mathematics, film, or any other context in which the term has previously been used.

Instead of explaining what we mean, then, it might be more expedient to deploy an allegory. If power is a bird-eating spider, then each of us is a baby kestrel. That is to say, those who dare emit a peep in front of the wrong billionaire's skybox will find themselves vigorously clamped between the chelicerae of the state security apparatus. If, say, two humble professors were to approach the managers of an electronics store, and politely inform said managers that said electronics store was henceforth to be the property of the commonwealth, run in the interests of the workingman rather than those of the investors of the Northwest Florida subfranchise of the RadioShack Corporation, what do you think would happen to these gallant champions of the meek against the merciless? If you believe they would immediately be handed the keys to the supply room, and apologized to for decades of irresponsible profiteering and shoddy consumer goods, you could not be less on-the-money.[125] Rather, as the astute predictor of power's machinations will already have surmised, said professors would more likely find themselves spending an uncomfortable evening in the dingy confines of the Pensacola Beach detention facility.

How then, is power to be tamed? *It does not take a lion tamer to tame a lion.*[126] But it does take *courage*, and courage is precisely what today's would-be-anticapitalist, pseudo-anti-lion crowd are dismally lacking in.

DECENTRALIZATIONISM

The problem may not be that power is too powerful. The problem may rather be that power has congealed into a few enormous gobs, each of which clogs the highway to utopia. But *clogs are for Dutchmen*, thus power must be decentralized.

Consider the authority wielded by the Chief Executive Officer. He may descend to the shopfloor and harass the technicians; if he attends a play he may flog the understudies. He may purchase all of the houses around yours and flatten them, robbing you of your one joy in life, the annual neighborhood block party and the opportunity it affords you to simultaneously snarf frankfurthers and proselytize your theory of geodesic banking to every resident of the cul-de-sac.

---

125   Remember: *"Caviar is the hubris of the mild."* Aphorism #31.

126   Aphorism #32.

But if our CEO is still unsatisfied, he may wander the streets handing out indecent proposals like candyfloss. The power of his mountain of silly little papers is such that he could ask the poor and sick to kneel before him in exchange for treatment money, or offer tramps large wads of bills in exchange for their performance of undignified interpretive dances and the grateful cooing of his name.

Is this kind of godly power one of the tolerable tyrannies? Is it indisputably aboveboard for an executive to carry the magic wand of money about the world, cajoling peasants as he pleases? If this man buys up all the vaccines and destroys them for pleasure, will we defer to the free market when the next outbreak of walrus flu wreaks mayhem upon our land? To answer affirmatively is, we posit, simultaneously to answer monstrously.

The redistribution of wealth is a decentralization of power; the dollar bills flow not just from rich to poor, but from center to periphery. The chief concern is that nobody wield more power than he can carry home in the pocket of his smoking jacket or romper. By all means, give each individual a little patch of shoreline over which he can reign with glee, but to allow a superpowerful oily oligarch like a toxic Jobs or a pungent Musk to buy up the whole coast and begin selling off timeshares is very much against the spirit of the thing.

For a small bevy of billionaires to be able to decide "I would like to buy a series of aeroplanes today" when your ordinary vagrant can do nothing of the kind suggests a social order that has descended into such arbitrariness that it has become absurd. Possession of the power to purchase planes ought to reside equally within each human, whether or not their pedigree is spotless or their conception was a mistake.

Longtime readers will know that we gallantly affirm the right of workers to own their factories, maids to own their hotels, waiters their restaurants, watchmakers their watches, flight attendants their skyplanes, journalists their blogs, and animals their zoos. Miners will mind their mines, the word "mine" at last ceasing to be a sick joke about property allocations under capitalism and instead becoming an accurate descriptor of the miner's true relationship with his hole. The only exception to this general rule concerns Surly DMV Ladies. Surly DMV Ladies will not own their DMVs. Surly DMV Ladies will be kindly retired to faraway pastures. There are to be no DMVs, which are incompossible with a sparkling tomorrow.

The Randomly Selected Congress

What could be more conducive to democratic expediency than the Randomly Selected Congress? What could more perfectly simultaneously embody the twin national motivating principles of egalitarianism and arbitrariness? And yet here we are, simply running and re-running the old electoral seltzer machine, in which the sociopaths inevitably bubble to the top.

Here is how the scheme works: Each American (excepting Floridians) will be given a slip of folded paper. This paper will have either a dot or a dash on it. At the appointed time, all will be instructed to unfold their slips. Those with dashes become Congresspersons and are to serve two-year terms. Those with dots may go off and sulk, and may be given a consolation cookie if necessary.

"But," readers and political scientists will surely note, "would not the results of such a scheme be disastrous?" Yet the questioner, though correct, has made a mistake. For the question here is not whether a randomly selected congress would be disastrous. It would. The question is whether a randomly selected congress would be *worse*. We posit that it not only would not be, but could not be.

As the Capitol's hallways fill with slightly dazed shopkeepers, mechanics, petty criminals, and mule-skinners, each newly drawn from Democracy's fickle hat, we are confident that civic-spiritedness will receive a much-needed shot of espresso in the arm, and that C-SPAN will reach hitherto unprecedented levels of watchability.

The Forest Congress

One can select a congress in manners other than randomly. One can also select it arboreally.

In probing applications of this principle, which ponders whether trees might govern better than men, we might eventually reach the end advocated by Keats, in which:

> *"every human might become great, and Humanity instead of being a wide heath of Furze and Briars with here and there a remote Oak or Pine, would become a grand democracy of Forest Trees!"*

Keats appears profound,[127] but his proposal comes several centuries too late, as we have already proposed a "Forest Congress," which would share power with an Executive consisting of Mankind and a Judiciary consisting of the Seas and Stars.

Here is our humble and decent proposal: Let us put the woodlands in charge. If we trust that Mother Nature is better acquainted with the moral markings of good stewardship than we, then let us surrender our authority to the Forest Creature Councils that have been increasing in popularity so rapidly during the last decade. By allowing the Stoats and Beavers to decide matters of grave scientific importance, we delegate our most difficult function and save ourselves significant lumps of time. Yet we surrender none of our power, as we are the ultimate carriers-out of the decided-upon tasks. It is a simple division of responsibility into a leafy Legislative branch and fleshy Executive body. The wildlife makes the laws, we execute them. We are the Presidents of Nature, a title which should suitably satiate our egos without giving us the unchecked authority to go on a bloodthirsty power-frenzy of logging thousand-year-old trees to make novelty greeting cards and pouring crude oil over pelicans just to watch them flail. No, no, the Beavers are our Congress now, a system almost scientific in its perfection, given the strong resemblance Beavers bear to Congressmen.

THE NON-EXISTENT CONGRESS

But through careful flexing of the imaginatory muscle, one can even endeavor to picture a thirdmost congress even superior. For what congress stands above all others, save the congress that has so successfully executed its mandate as to eliminate the necessity of its own existence? The Forest Congress is but a leafy pitstop along the arborway to the Ideal.

Congresses get in the way of reason, and always have. The only reason our legislative sessions are kept so short, and our legislators sent off to their faraway districts for permanent campaigns, is that this is the best way we have found to keep them from voting for anything. But if that is the case, it would seem as if an even better

---

127 A convincing scholarly case has been made that there is a direct correlation between the stylistic techniques a person uses to draw a tree and that person's definitive psychological makeup. See K. Bolander, *Assessing Personality Through Tree Drawings* (Basic Books, 1977)

situation would be to eliminate congresspeople altogether, instead of constantly having to paralyze and misdirect them in order to keep rumpus from occurring.

Send power outward and westward, back from whence it came. Banish it from the Columbian District, where it has done nothing except pollute a once-beautiful city with fumes and legislation. Never let us succumb to the belief that governments are necessary.

Why, listen to the words of one of the captives themselves, describing in his memoirs the achievements of his chamber:

> *People ask me, looking back over a four-decade legislative career, what I am most proud of. I've named a lot of post offices, re-allocated a lot of highway funds. I once slipped a rude haiku into a last-minute spending amendment, which is still technically the law of the land! I funded the construction of a very large plane, which went missing. And so no, I don't feel as if it was a waste of time.*[128]

Has self-obliviousness ever apexed more entirely? Will not this brief excerpt furnish an adequate exhibit to summarily conclude our trial on the question of whether a Congress ought to be allowed to persist? Ladies and gentlemen, we rest our case and gather our things.

We can randomly select our Congress. We can turn it into trees. But nothing sparkles brighter than a Congress so translucent that it ceases to exist altogether.

WHAT THE POLICE SHOULD DO

> *Find happiness through dressing like a policeman*
> - Slovak proverb (alleged)

> *"We knew there would be a secret police,*
> *but we didn't know it would be such a secret police!"*
> - Arthur Miller, <u>The Interrogator</u> (1947)

But once we get rid of the Congress, what of their enforcers? Are police inevitably also obsolete? No. The cynic replies that it is time to snip the blue line altogether; if not exactly to remove it an elabo-

---

[128] Rep. Charles M. "Chili" Murphy (D-NM), *Speaker of the Horse: Reflections of a Cowboy Congressman* (Vanity Press, 1998), p. 822.

rate anarchist bris. But if there is anything to fear more than millions of policemen, it is millions of out-of-work policemen with countless free hours to roam unaccountably, and no further paperwork to keep them distracted. This practical side of things leads us to oppose proposals to immediately disband all local and state police forces.

No, we believe the police can be put to good use, if not exactly redeemed. Perhaps their mission could be subtly readjusted, toward something socially useful instead of the current program of distributing free and low-cost bullet wounds to disadvantaged black youths. While that present program has resoundingly achieved its objectives, we cannot help feel as if its benefits have been dubious.

There are ways to repair even a police department shattered into a thousand wayward moral shards. With today's reconstructive technologies it would be entirely possible to put Humpty Dumpty together again. Just because we fear the police does not mean we should give up all hope of someday sitting down to breakfast with them and laughing together about the regrettable triviality of our prior misunderstandings.

But for the police to achieve any worth, they will have to be adjusted mightily indeed. The first necessity will be the adoption of a new aesthetic. Tonsures are a must. The regimental tonsure will go a long way to reducing the policeman's credibility with the populace, essential if the cop is to be seen as a coequal with the criminal. The elimination of unnecessary glinting baubles, such as badges and firearms, will also form a core part of the makeover. Uniforms will be of a deep, rich magenta, the better to convey a sense of flair and fancy. Neckties, however, will remain a dark black, as a somber reminder of the death inflicted by police during more ominous times.

But adjusting the color of a uniform only adjusts the man within to a limited extent. It will still be necessary to repoint the tracks toward an alternate horizon. The police must not simply be trying to look better, but to *be* better.

To this end, their function should emphasize the "serve" in "protect and serve." A policeman will leap in front of a screaming bullet for you, yes, but he will also fetch you an ice cream upon request, or draw you a picture of a sad-eyed duckling. She will not shy from the task of assisting bank robberies, but she will also not let that get in the way of painting youngsters' toenails. She will not always smile, but she would never hit you. *A policewoman's presence in one's neighborhood will be a comradeship rather than an occupation.*

Extremists often carry the sway of discourse, accustomed as they are to getting their way. But the radical anti-police abolitionists are, at least as regards police, wrong to a fault. We can save our police forces, but we must direct them toward the offering of complimentary brownies into the hungry mass instead of the pumping of complimentary ammunition into the spinal column.

COMFORTABLE PENAL CONDITIONS

THE IDEA OF A JUMPSUIT tickled the provost. "No, Mr. Harper," he chuckled, "this is not a wearing-stripes-and-breaking-rocks kind of penal institution. Our methods are gentle encouragement and comfortable bedding. But it's not all boule, either."

- from <u>The Prison in 2020: A Speculative Tale</u>,
Federal Bureau of Prisons Printing Office, 1940.

If it truly is the *sparkling* with which we are concerned, we must face up to the fact that our prisons do not sparkle. They are, by contrast, worryingly matte. Many of them are downright dismal things, with little thought given to color schemes or proper feng shui.

Most of us have not enjoyed our time in prison. But the serious question is whether it had to be this way. Couldn't a stretch in the old incarceratorium have been, if not fun, at the very least *relaxing?* More like a sauna than a torment?

"Incentivity!" the sadist cries in response. "One must have the proper incentives in order to act properly. If prison is a bounty, its canteen serving luxury jellies on reformatory toast, what becomes of the discouragement factor?" The question could not be phrased more sensibly. It is nevertheless wrong.

It should be noted that the one prominent criminologist upon whom we forced this idea commented that it was "very interesting but hardly feasible." But this discouragement only encouraged us. "Interesting!" Bah! The word "interesting" is a clever dodge, a trenchcoat worn by seedy ideas to disguise their base intentions. We all know what "interested" the bishop, *thank you very much.*

Incentives are surely not to be jousted with, but it is not impossible for someone to win a joust. But here is our truth: a pleasant prison has no chance of attracting the throngs, for even though the latrine-seats in the Prison of Tomorrow are flounced and tufted, *inmates must nevertheless clean latrines.* It provides dignity and comfort without complacency and sloth. It is carefully oriented toward the preparation of the human for life on the outside world. See the provost's remarks on boule, above.

We firmly believe it is possible to reform the deviant without subjecting him to humiliation and disgorgement. There is no reason that a prison jumpsuit cannot be complemented with a stately cravat. "Gaols into Bouncy Castles!" must be our mantra, and let us never forget it.

Or Perhaps No Penal Conditions Whatsoever

But there could be a faulty premise. Perhaps there is in fact a time to think thoughts that are radical rather than sensible. We should think on the grand scale, as it were, as the futurist projectionists of the past had no hesitation about doing. Edmund Wilson's *To The Crumbling Breadhouse* describes the man who, sneaking through a curtain, saw human existence without her brassiere and immediately fainted. Yet how long has it been since our own generation had a thought so profound that it immediately fell unconscious?

In the season for boldness, then, one must sprinkle bolder seasonings. And where prisons are concerned, it may be time to think not simply "How can we ameliorate this?" but "How can we rid ourselves of this once and for all?" Now, certainly it is difficult to conceive of a world in which a certain percentage of the populace isn't stored permanently in crates for their own good. But we dare to posit the extraordinary.

Prison has always been nice for a visit, but who would really want to live there? And if we recognize that none of us truly would, then surely we have also recognize that we ought to be fashioning a way to unimprison ourselves, to whittle the shiv that will scrape our way to liberation.

"If we shuttered all the prisons, what would we do with all the prisoners?" a child might ask. Well, small lass or laddie, perhaps we could give the dangerous ones an island and just keep a good eye on the others. Or perhaps the child has a question with a mistaken premise, and we should be bothering ourselves more with questions of how to prevent persons from becoming the persons that then become prisoners to begin with, rather than where to place those persons subsequent to their so becoming.

It will be difficult to let all of the prisoners out at once, if only because they will have a difficult time squeezing through the door. But the goal must always be borne in mind: get the inmates out the door at last and set the facility aflame.

WORKER OWNERSHIP

But we will not have eliminated prisons until we eliminate the workplace. Let us consider one of its various injustices.

Think of the workplace's governance arrangement. Did you get to vote for your boss? Did your nephewess? Did Ricky? Yet is not it an elementary principle of the *Democaractacus* that one's affairs should be governed by oneself? What explains the disjunction here? In a sensible society, would not one's bosses all be elected?

Yet somehow this is not only implausible but veritably risible. Say that you and your co-toilsmen approached the foreman or overseer one dewy morning and said the following:

> We employees recognize that you are in charge around here. What's more, we are in general agreement with you on first principles, as we ourselves are capitalists. (We believe in having things and in owning them.) However, we have decided to vote on you, as seems only fair. There is to be an election at noon. You may run, collect donations, and speak persuasively. You may yourself have a vote, as well. Upon the election's completion, as is the usual process, the victor will be anointed with boss-juice and given the post.

It is sure as shinwater that nothing good would come of this.

Perhaps it is thought by the cretin that human beings are only quasi-capable of sorting themselves into bunches and deciding when they are ripe for plucking. We would not trust the banana with a similar duty. But the human being *is not a banana*. When fruits produce their Wittgenstein, the comparison will self-validate, but until then we feel it can be dismissed *with prejudice*.

Nevertheless, there are consequences to absolute powerlessness. After all, have you ever stopped to ask yourself: "Why don't I get to take a rope-bridge to work?" The answer is almost certainly "because it is not permitted." But who permits the permittor to permit? Why, the people permit the permittor, of course. But what if the opposite were true?

In posing thousands of similar rhetorical queries, we uncover a number of important messages regarding hierarchical management structures and workplace democracy. For example: is it any coincidence that the words "manger" and "manager" are only a letter apart? Probably not, but this in itself makes the manager neither a French restaurant nor a crypto-Christ.

The more vital lesson is this: no system of management can be said to be "sane," or at least *reliably* sane. Because every human is vulnerable to the same impulsive desire for education and edification, we are each equally liable to suddenly lose our minds, thereby rendering the risk of allocating authority to any one individual far higher than any person *with* remaining sanity ought to be willing to gamble upon.

So: all workplaces must be owned and operated by the workers, lest managers succumb to sudden madness. To be clear: this is not advocacy of the cold, uniformed tyranno-communism your textbooks warned you of. This is a joyous, celebratory workplace collectivism, in which all persons whistle tunes and tell jokes while selling the books or peeling the rinds from the floor.

Therefore: ask your boss not for a raise, but an election.

Tiny Landlords You Can Squash

The boss is not life's only *jefe*. Each place one goes, one is instructed and cajoled by the authorities, who mutate into various forms and fora. Here the supervisor tells us to buckle up and quit slacking, there the tyrannical campus parking attendant informs us that the prohibition on junior faculty using spaces designated for senior faculty will be strictly and pointlessly upheld. Into each of life's junctures small despots insert themselves.

Yet this descriptor is precisely the problem, isn't it? For to characterize these despots as small understates the enormity of their wrongdoing. The issue is that these despots are *in fact quite large*, they thunder and stamp across every pasture and gulley.

But before returning to the abovemost thought, let us pause ourselves to consider the evils of one particular giant shorecrab: the landlord. At the initial soft blush, landlords do not make sense. Why should I pay a sum to the rentman who then pays a smaller sum to the mortgageman, simply because he possessed a larger sum to begin with? Why does he get to keep the house at the end, while I am to count myself among the elect if I even manage to have my security deposit returned? Why all of this in spite of the fact that he enters my quarters when I am absent, rifling through my knicker-drawers for evidence of tomfoolery and horse-love? In a world built on sense instead of dollars, surely proposals for the existence of such a man would be quickly jeered away, dismissed as the product of some sinister mental defect. What next, a man who controls your job and shouts at you about it without himself being able to do it or ever having even barely understood it?

No, landlordism is an unnecessary idiocy, especially with the onset of the Internet. Why, would it not be a simple matter to use some sort of website to list all homes not currently being lived in, and allow the individual to choose which one she would like to have the state lend her at no charge? Ensure every home is lived in, do not let plutocrats amass large quantities of useless country manors, and freely distribute the stock to the citizenry. What could be lovelier?

But if there is one thing we are about, it is the recognition of practical constraints to the implementation of our proposals. And where free luxury housing for all is concerned, we are aware that our carefully laid plans face a man-sized difficulty, namely that landlords are physical human people who cannot simply be tossed in the sea. It's all well and good to go spewing calumnies against these profiteering devils, but are they not equal to us in the flesh, possessing lungs and a nose like the rest of us mortal fools? With the exception of Noseless Joe, the proprietor of the shady, tick-ridden flophouse in which we temporarily resided during a low point in our graduate school years, the answer is yes.

We may not wish to exterminate the landlord, then, but to make him somewhat tinier. Yes, yes, by all means free houses, that is a given. But in the meantime we can at least adopt a doable moderate reform: make sure no landlord exceeds the size of a raisin, so that if he becomes insolent or refuses to repair a utility duct or eradicate a Kafkaesque bedbug, he may be instantly squashed in the palm. Thus is the power dynamic subtly readjusted.

RECLAIMING THE AMERICAN DREAM

The subtitle of this book, *Thoughts on Reclaiming the American Dream*, is no accident. It is also the subtitle given by Barack Obama to his own book, *The Audacity of Hope*.[129] By gently lifting it and applying it to our own work, we hoped to catch a ride upon the coattails of his Google search-results and, God willing, cause numerous accidental purchases. The substance of the words was of far less import to us than their effect in deluding the consumer. It will be left to historians to determine whether we were successful in fulfilling this intention.

But for us, the American Dream is something more than just a devious thieving scheme to defraud the public. In this text, we have indeed made an effort to reclaim it from its usurpers, and in in our attempt to do so have applied the entire family of powerful thought tools handed down by the great thinkers of the age.[130] We have used everything from *topology* and *geodesics* to *general systems theory* in our attempt to show you a *true* American Dream, one based in neighborliness and hammocks rather than fence-based economic delusion. In providing practical proposals rather than platitudes, we believe we have done more real work toward the preservation of the American Dream than any contemporary Obama.

But let us additionally consider the words of Dr. Freud, who with his most serious face describes the dream as "the substitute for the infantile scene modified by transference to recent material." The American Dream is therefore a manifestation of infantile experience, which has "seized the opportunity offered by the continued cathexis of painful day-residues, has lent them its support, and has thus made them capable of being dreamed."

Operating from this psychological foundation, we may reasonably attribute the American Dream to nothing more than a collective repressed neurosis, and the issue of reclamation may therefore be swiftly disposed of.

---

129 See Barack Obama, *The Audacity of Hope: Thoughts on Reclaiming the American Dream* (Three Rivers Press, 2006)

130 Buckminster Fuller.

THE END

Perhaps it is true, as a certain *New Yorker* critic has alleged, that we are guilty of nothing more than "goofy futurism and a boyish love of dinosaurs." Surely that does not in itself discredit the Points, though, or mean that a series of two or more Points does not create a Line from here to Tomorrow.

A choice has set itself upon the land.

Do we wish for peasantry or wonder?

Pastries or demise?

Hegemony or survival?

UTOPIA OR OBLIVION?

Sketching a Blueprint with Wobbly Hand

The sad general reality is that, too often, our finest citizens loaf around spouting aphorisms and drawing blueprints, when the necessity is for ACTION! Certainly, our hearts can be cheered with a song, but a song will find it difficult to construct a feed-barn or apply a new coat of paint to a community milk-van. We ourselves have been somewhat unfairly accused by several critics[131] of being mere pontificators, content to issue pronouncements and proclamations from our Endowed Thrones at the tippy-top of Ivory Tower University.

This section addresses the concerns put forth by Messrs. Cavett and Krauthammer,[132] by getting straight down to business. While we cannot yet entirely overcome the theory-practice gap in book format, as any book without limbs or a number of knobs and dials is destined to fall onto the "theory" side in the chasm, we can become as specific as possible with our proposals. Much of the slander that has been hurled at us from syndicated columns and mansard rooftops has posited the same core query: "What would you have the Society actually *do* in the Here and Now to implement your notions and musings? What Federal Legislation must be passed?" While we reject this criticism, we acknowledge it, and wish to forestall its furtherance by stating clearly, luminously, and precisely what we would have The State do, should Reason suddenly begin to govern her mandible.

Therefore, for those wretches who demand them, here in glittering specificity are our proposed Constitutional Amendments:

> ***Amendment 29:*** *Each person shall be guaranteed a small park within walking distance of her home.* This would encourage leg-stretching and would give children an enjoyable place to frolic during the summers. Benches should be present within.

> ***Amendment 30:*** *There shall be a 100% tax on inheritance.* During our time in the Academy, we have encountered a number of incurable Nincompoops and

---

131 Most notably Charles Krauthammer of *The Washington Post* and Dick Cavett of *The New York Times*.

132 See previous footnote.

Dunces who have managed to reach levels of modest to extreme success purely by dint of the hard work and/or brains of a parent, grandparent, or even more distant ancestor. If we are to achieve a true Capitalistic Meritocracy, each must start out in the exact same place as the other.

Amendment 31: *Politeness is recommended, but not required.* We do not support mandatory politeness, for it may prove bothersome. We do believe that a constitutional amendment could function as a more forceful variation on the Non-Binding Joint Resolution, demonstrating a general sentiment shared by learned people without being overly fascistic.

Amendment 32: *No horses.* Horses are a blight. Dispose of them forthwith![133]

It may be noted that we have not proposed a twenty-eighth amendment. We are operating on the assumption that the Equal Rights Amendment really ought to be a prerequisite for amendments about politeness and public parks.

But, look: do we wish for Progress? Of course, we detest *Progressivism*, for we believe that Progress for the sake of its own self is deeply senseless. Yet we similarly spurn attempts of Stagflation, Regression, and Traditionalism to curry our affections and affect our curries. Instead of valuing Tradition or Progress, we must value values.

We believe that finding the Human Values and measuring our success or failure at matching our realities to them is the ultimate test of whether society has progressed towards death or regressed towards infantilism. We also believe that these Human Values can be quantified in terms of the temperature of human actions. Cold human actions denote a regress toward murderousness and social mistrust, while warm human actions represent lovingness, satisfying meals, and

---

133 This Amendment naturally exempts the horses used by the famed Mounties of the Canadian Isle, whose ethos, fortitude, and sonorousness are still much appreciated by the authors, and who have proven themselves to be generous and understanding when it comes to requests to quietly hush up ugly incidents of alleged felonious moose-impersonation that could have jeopardized a promising (though ultimately fruitless and infuriating) job interview for a Visiting Assistant Partial Lecturer position at the University of Winnipeg.

social connectedness. The frigid antithesis to human freedom is the Gulag, while its warm brother is the tropical Jazz Festival. Not for nothing did a clever bastard once say that the aim of his movement was to "build a real-life Buffalo right here in Louisiana."

Progress will be therefore measured not in terms of our efforts to inhibit global warming, but our efforts to create global warmth. Far more vital than any Prescription or Program is the emergence of a general sense of coziness and pleasure on the part of the populace. The most effective and generous anti-poverty bill in the history of Congressional Legislation will be abysmally catastrophic if it is Cold and Bureaucratic. By contrast, the most meager and fruitless local efforts to achieve dignity and prosperity may be considered great successes if they manage to increase the overall amount of Pleasing Warmth in the community.

Therefore: Construct new wombs! Make us feel Comfortable, Secure, and Warm!

But wombs and womb-like substances are only the beginning for our Community of the Marvelous. So let us take these moments to map our Utopia, so as to provide the most useful possible guiding tools for the Generations to Follow. Wombs aside, vital elements include:

- Safe, efficient, and non-aggravating public transit
- The replacement of newspapers with wandering calypso singers
- A high-quality university that gives no Grades and accepts all who wish to learn within its walls
- Tree-houses and rope bridges!
- A downtown movie parlor with reasonably-priced tickets and an endearing elderly person who works the concession stand
- Free rental of automobiles, zeppelins, and bicycles
- Cobblestones
- Tea rooms and coffeehouses
- Begonias abounding
- Plenty of parking, though not to the detriment of walkways and shady-places
- Hammocks aplenty
- Meetinghouse for democratic decision-making

- Tail fins on every motor vehicle
- A ban on the filling out of forms
- Free musical lessons in the gazebo
- A gazebo in the park
- A park in the middle of things
- Swan boats on the water, inner tubes in rougher areas
- Useful police who hand out cocoa on streetcorners rather than busting the unsuspecting
- All golf is to be miniature-golf
- Reusable packaging for all food and beverages, lots of glass bottles with emblazoned logos
- Milk bars, juke joints, and honky tonks
- Free problem-solving offices, tasked with assisting persons with the solving of all conundra and entanglements they may face
- Enormous public libraries
- Monthly community BBQs
- A labor system whereby:
- All goods and services are available free to residents, at cost-covering fees to tourists
- Work for a certain minimum number of hours maintains free access
- Required hours drop for unpleasant work
- Extra work earns great respect
- Profit is neither sought nor found
- Ubiquitous water fountains
- A forest congress
- High quality confectionary shops and bakeries
- Tunnels and sky-bridges for the cold weather
- Useful maps and well-marked streets, but also secret gardens
- Fifty different cheeses (**and no more**)
- A bench on every sidewalk
- Rapturous respect for those who teach

With these necessities enumerated, readers may commence the transformation of existing shopping centres, business parks, and university campuses. Begin by placing a flamingo in the dean's office and stealing this book.

CONCLUSION

But all the psychodrama and motoramas aside, let us offer some tranquil reflections:

We initiated our ramblings with an excerpt from a particularly horrifying news story, in which students were being continuously threatened with death for the mere act of performing ceaseless hugs upon one another. Here at the end, we offer you another startling testament to the times, this time from Atlanta's WXIA News Channel 11, as reported in March of 2010:

> *Lima, the zebra that escaped from Ringling Bros. Circus and wandered onto the Downtown Connector a few weeks ago,* **has been euthanized.**

We believe that even the most hardened and frigid Economists among you, dear readers, will agree that the levels of zebra-euthanization in this country are intolerably high. Not only that, but the fact that a national month of mourning is not initiated after each execution (zebra and otherwise) suggests a society in which mass sociopathy has seized the reins on the stagecoach of power.

Yet there is reason for optimism! As miserable as conditions for zebra and man alike may appear, the present is ripe with the future! We find hope in the innate human longing for Paradise, the Moutaintop, A More Perfect Union, the Halcyon Days, the Revolution, the City on a Hill, Progress, for Total Liberation of The Mind, for Orgasm, for Dynamic Equilibrium, the New American Century, for Perfection, for the Classless Society, the Meaning, for the Greener Grass, for a world where *even though there will be magicians, nobody will have any illusions.* This striving towards a nonexistent endpoint, this "belief in an attainable paradise," drives all forward motion.

Motion without direction is chaos, however, and chaos is as far from our intention as the ant is from the ant*eater*. We have a guiding principle, and it is this: The Uncompromising Coital Embrace of the Absurd.

This does not simply mean Being Silly. It is the inverse: a serious grappling with the vast unfeeling pointlessness of "the" Universe, and an attempt to find meaning *within* recognized futility. Here we find inspiration from our mentor Mr. Camus. In 1942, Mr. Camus published *The Myth of Sisyphus,* in which the human condition was portrayed as that of Sisyphus, condemned forever to push an enormous boulder up a hill, only to have it roll back to the foot each time it broached the summit. Mr. Camus suggested that Life itself was equally devoid of consequence or possibility. Yet all was not lost. Against all reason or expectation, young Sisyphus can find meaning in his struggle. Each time he reaches the tippy-top, as he watches his boulder begin to slide once more, he may shout from his hilltop: "To hell with the cabbage-headed bastards that put me here! I will roll boulders for eternity, and what's more, I'll enjoy it!"

Unfortunately for Mr. Camus, his little wind-up philosophy box seemed ill-equipped to deal with the main intellectual horror of his age: Hitler. Our little planet was at that moment in its Crisis of Values, in which the Meaninglessness and Valuelessness Mr. Camus had said we could enjoy appeared to lead in the direction of Relativism and subsequently Totalitarianism. Or at least, such is the half-remembered story. What were we to do?

Mr. Camus had the answer once again. We would moderate the Meaninglessness with Humanism! We may not be able to find God or Eternal Truth or The Pot O'Gold, but as long as we were stuck on Spaceship Earth together we might as well enjoy one another's company. In his *Letters to a German Friend,* our Camus wrote of his dismay at the Nazi conclusion that Nihilism was the proper belief to be derived from Absurdity. "That's not what I meant!" one could hear him cry out amid the wartime darkness:

> *What is truth, you used to ask? To be sure, but at least we know what falsehood is; that is just what you have taught us. What is spirit? We know its contrary, which is murder. What is man? There I stop you, for we know. Man is that force which ultimately cancels all tyrants and gods. He is the force of evidence. Human evidence is what we must preserve. ... If nothing had any meaning, you would be right. But there is something that still has meaning. ... You never believed in the meaning of this world, and you therefore deduced the idea that everything was equivalent and that good and evil could be defined according to one's wishes. You supposed that in the absence of any human or divine code the only values were those of the animal world — in other words, violence and cunning. Hence you concluded that man was negligible and that his soul could be killed, that in the maddest of histories*

the only pursuit for the individual was the adventure of power and his own morality, the realism of conquests. And, to tell the truth, I, believing I thought as you did, saw no valid argument to answer you except a fierce love of justice which, after all, seemed to me as unreasonable as the most sudden passion. Where lay the difference? Simply that you readily accepted despair and I never yielded to it. Simply that you saw the injustice of our condition to the point of being willing to add to it, whereas it seemed to me that man must exalt justice in order to fight against eternal injustice, create happiness in order to protest against the universe of unhappiness. ...I continue to believe that this world has no ultimate meaning. But I know that something in it has a meaning and that is man, because he is the only creature to insist on having one. This world has at least the truth of man, and our task is to provide its justification against fate itself. And it has no justification but man; hence he must be saved if we want to save the idea we have of life...

No, friends, we say that Nihilism has no place, in spite of its tantalizing ease and the handsomeness of certain leather garments. We must have an Existential Humanism, an ideology that recognizes both the futility of our strivings and the indispensability of our lovings!

But Universal Love cannot simply be spewed from the pulpits or rise from the sea. We must stress a unity of Theory and Praxis if *Blueprints for a Sparkling Tomorrow* is to have the impact that its rear cover promises. Back-pattings and the occasional high-five do not constitute sufficient fulfillment of the doctrine of the eternal brotherhood of all mankind.

What is to be done? Allegiance to the aforestated Aphorisms is, of course, necessary and proper. But our direct prescriptive edicts, sage as they may be, can only carry our fellow terrestrial-cosmonauts so far. Long-term adaptation to one's environs requires a dynamic process, which changes with the times and provides the individual with the answers to each query that may arise, no matter how many future-technologies and hitherto-unforeseen-eventualities it may involve.

Thus, we here expound an ethic: universal anarcho-amorousness. Love as sole law! Replace the carnivorousness of plants and man-beasts with a rigid attachment to the interconnectedness and inter-amorousity of beings. To this end, we believe that each and every statute, regulation, and ordinance must be repealed forthwith, and the Founding Documents of our Great Nation seized and set aflame. What has the Privileges and Immunities Clause to do with the daily acts of today's factory worker or haberdasher's apprentice? All of us have a fondness for speedily-delivered letters and telegrams, but surely Government in all of its non-postal functions is an impediment rather than an enabler of *Virtù*.

Also, pocketwatches. We support their repopularization. The only route to liberation is through mandatory universalization of the Gentleman's Timepiece.[134] If our age can be said to have a hero, it is most assuredly Vincent P. Falk,[135] yet we have thus far failed to deploy his model on a grand scale.

As a further recommendation, the need for a New Plague of Frogs has been growing for over two decades, and we believe it is finally time to travel that gleaming airport moving-walkway that stretches from Thought to Action.

Will our tomorrow sparkle? Will it be awash in the neon-pink glow of a warm train-car diner, or will it suffocate beneath the noxious stench of a day-old possum carcass? Only Tomorrow can know. But Tomorrow never knows! How then, can a prediction be made, without resorting to gross speculation and conjecture? The fault of this question is in its pejorative use of the terms "speculation" and "conjecture." We believe these are terms to be embraced rather than derided, and that if they are properly saddled and trained, they may be employed in the service of humanity rather than senselessly persecuted.

The predictive profession is perhaps not the wisest pursuit for two earnest young scholars such as ourselves. We do not wish to end up like Whisenant or Erlich. Yet the predictive power is a persuasive power. By forecasting, we simultaneously create. This is precisely the reason our tome is considered a *blueprint* rather than an *etching*. These potentialities can only be realized through the employment of the blueprints themselves. Laze and malaise will produce no Motel of Tomorrow or Shimmering Serpentine Futuropolis. They will not bring us Tomorrow's Oven Today and will neither Reshape the Bagel nor Dissolve the Unitary.

---

134 This contradicts, of course, a prior section positing time itself as a corporate conspiracy and proposing to replace it with some kind of heartbeat-rhythm. Please disregard that portion of the text, which was written under the corrosive influences of Wine and Women.

135 See R. Ebert, "Vincent P. Falk and His Amazing Technicolour Dream Coats," June 4, 2009.

What is Progress? When is Tomorrow?

These questions are among Man's most fundamental and unanswerable. Yet were they to be considered simplistically rather than philosophically, we might find the solution sought. Tomorrow comes after today, and progress comes from prestidigitation.

How long can our little planetary experiment sustain itself without plunging into the sea? Two weeks? One million weeks? An eon? A crypto-eon?

Our answer: none of the aforementioned! The answer is, in fact, As Long As We Keep Clean Noses And Heed The Warnings. No heed, no shoes, no planet! What heed? What warnings? The warnings offered herein over the course of multiple hundred shimmering pages. The essential distillation of those warnings reads as follows: without a well-defined Utopian Prospect, humankind is doomed to meander aimlessly down hallways staffed with Econometricians and Bureaucrats. As horses were to Melville, a solid Utopia is to the human. So, map the Utopia and then scour the map for the on-ramp that will reach it with the most pleasing balance of efficiency and aesthetic bliss. Futility matters very little, and should not dampen one's enthusiasm for the Quest. Begin immediately, and do not tire or exit until each aspect of social livingry has been either perfected or rendered tolerable.

Anyway, our Blueprints have now reached their termination point. The lessons have hopefully been absorbed. Apply as directed. Adjust as necessary. Re-read and influence friends.

We have completed our work and send you fond farewells. You may now lower the observation deck.

Epilogue:
Stray Molecules & Ancillary Corollaries

We must here address the failings of our own work, which are few but significant. It may be intelligently argued that certain sections of the preceding work are a bit too post-modern for their own good. This applies particularly to one particularly troublesome chapter in which saints, accordionists, and muralists appear side by side in a stew of meaningless word-matter.

We acknowledge the existence of this criticism, but do not acknowledge its veracity. Our work is far from the valueless Deweyan hodge-podge it may appear to the dogmatist. We do not attempt to negate Logic, Rationality, and Virtue with our text, although we may have done so incidentally. We do not intend to stick a finger in the eye of Allan Bloom or Bill Bennett, although if this kind of happenstance befalls us, we will not shed a vast quantity of tears.

We are not the neo-Discordianists some have labeled us, although these are precisely the words a neo-Discordianist might write in his own defense.

We acknowledge that by acknowledging the flaws of our work, we have transformed it into a meta-work, and that subsequently by acknowledging the meta of the meta-work, we have transformed it into a meta-meta-work, and that by acknowledging *this* transformation, we have doomed ourselves to an eternal return of the same. We do not much give many damns about this.

Above all, in spite of appearances to the contrary, we do not believe in Nonsense for the sake of Nonsense. We believe in Nonsense for the sake of Progress, although not a lick of what we have said falls into the category of Nonsense, as anyone with an advanced degree in it could inform you.

We apologize for any further unrecognized flaws with our methodologies or conclusions. Much was removed from the book in the editing phase, including the majority of the important equations and evidence.

A word on authority: We know we are philosophers instead of kings. As such, we know our tiny wordlets will hardly be seen from aeroplanes or pecked at with telescopes. If we are one thing, it is modest. Nobody is delusional enough to believe that each word in this book is instantaneously translated into a binding public policy upon the moment of its utterance. However, it concerns us that our national gossip magazines tend to take it more seriously when a king leaves his house than when a taxi driver does. For whosoever may call himself a king is, in the eyes of himself, a king.

We recognize the slight disjointedness of our writing-style. This can be attributed to the fact that many of these writings first appeared in print as weekly *National Review* columns, or as essays and asides for *The Zagreb Review of Books*. Several chapters were originally given as speeches to the Commonwealth Club of Alameda during their annual Member Appreciation Luncheon, and two chapters are adapted from lectures given to the Ornithology Department of Brandeis University (formerly Middlesex Veterinary College). A certain level of information loss is thus to be expected thanks to these transfers between mediums.

At one point in the text, the reader was addressed as "comrade." Please do not take this as an indication of any particular political sympathies on our part. We have pledged allegiance to no clan or guild.[136]

It may also be said that we suffer from The Problem Of Simultaneous Embodiment And Parody, as well as a mild form of specicism. Each of these criticisms would be accurate, were we not operating on a meta level. However, as we are, the falsity and debauchery of these claims is not only demonstrable, but definitive.

In direct and utter sobriety, do consider the following: Ninety-Five percent of what has been demonstrated here is directly taken from Reality. The Absurdity has been embellished, but never invented. The world portrayed here by us is indeed the world both you and we inhabit. Enjoy.

---

136 This renunciation of politics is not absolute. We have resisted pressure to rescind some of our stronger articles offering political interpretations of various fruits and fruit companies. This ranges from our staunch condemnation of the pulpier-than-thou attitudes of contemporary orange juices, to the harsh words we have expended on the banana for its historical rôle in rationalizing patriarchy. Yet we are not universally negative on the subject. For example, we have spent thousands of words honoring the grape, for the grape respects teamwork. *Grapes hang together.*

APPENDICES

Because we were committed to making the *Blueprints* a masterpiece of structure, a whole scout-troop's-worth of editorial whittling has been performed on the pages. "A book," our agent gently explained to us, "is generally rectangular." Alas, allegiance to convention triumphed, and our months of insistence that *if the book was not octagonal there was no sense in publishing it* gradually subsided. While we maintain committed to experimental geometries in nonfiction, we do occasionally stoop to allowing practicalities to invade our work.

In shaving edges off, then, we have been left with the intellectual equivalent of a barbershop-floor of hair-remnants. And if there is one thing known by barber and quartet alike about the hair on the floor, it is that it *cannot be reconstructed into a purchasable toupée.* If it could, many a hairdresser would today live in a state of hedonistic wealth and luxury. That they do not provides evidence for our theory.

The point, then, is that *one cannot reassemble the pieces* once the vase has been thrown against the tile by a shrewish wife who cannot understand that tenure committees are not interested in whether a man has devotedly prioritized his family over the completion of his *Journal of Housing Theory* article on the habitable capsule.

Once it falls off and shatters on the floor, a book's pages may as well be in the fire, is what we are trying to say. There are destined to be *remnants* and *leftovers* which, while not inferior to the final body text, have merely had the misfortune to be blunderingly severed by a simpleminded hack of an editor, of the sort who believes that airport potboilers and visionary academic treatises require equivalent editorial technique.

It is here in the appendices, then, that we place these "spare parts." We posit that they will have that acute usefulness for which the appendix is so widely known. In them, we shall sweep away the spare bits of carcass. We shall clarify some misinterpretations and ruthlessly disembowel our critics. These lectures, notes, and ephemera are therefore, not to be ignored. A proper understanding of the foregoing text cannot be had without properly understanding its appendices.

Appendix A

The Thirty-Three Aphorisms and Their Effect on Living

Fear of writing a new aphorism is fear of oneself.[137]

I. Living Aphoristically: Why Thirty-Three?

Throughout the main text, from Footnote Eight to Footnote One Hundred-and-Thirty-Seven, we have sprinkled notations of certain key apothegms, numbered from One to Thirty-Three. This is no accident. It has long been our contention that pithy mottos are both far easier to remember and far easier to produce than carefully-researched arguments. In preparing this book, we have therefore minimized the time spent cross-checking citations and copy-editing prose, and maximized the effort put into the emission of tiny sayings. In this way is the appearance of profundity most convincingly generated.

The aphorisms proposed above are, it will be noted, disparate and multifarious. Some come in the form of exclamatory declarations, others resemble queries or probings. This is a fully intentional effort on the part of the authors to stimulate multiple parts of the mind in rapid succession (specifically, parietal and occipital lobes). We are of the opinion that aphoristic sequences which create *rhythms* tend to breed numbness and incomprehension, and thus have opted to prioritize dissonance in our presentation.

We are far from the self-cented RANDians[138] that our profiteering book-publishing careers would indicate. We carry many classically "leftist" dogmas and smegmas around in our *attachés*. For example, we believe that all things should be held in common by the masses, and that the Captains of Industry should be either defenestrated or repurposed as sleeping-car porters. But we *do* believe the aphorisms should be guarded. Many a civilization has crumbled thanks to a misplaced aphorism or poisoned metaphor. Do not underestimate the power of the Word to incite the Deed!

---

137 Aphorism #33.

138 Research and Development Corporation, not Ayn, whom we hold in eternal contempt for her stigmatization of the shrug.

Yet do not slip into miserdom, alone in the woods hoarding stacks of aphorisms in an overflowing concrete filing cabinet. At their very crispy core, the 'phorisms are designed for Social Betterment, and she who forgets this truth forgets the aphorisms themselves. Simply use your judgment. Give them to friends at parties, but do not hand them out on streetcorners or sell them to fortune-cookie manufacturers. Careful employment of Wit and Good Sense will guarantee proper usage and a future filled with aphoristic bliss for all.

II. Employing the Thirty-Three Aphorisms in Your Day-to-Day Activities

It's Sunday morning! When you go down to breakfast, shout a few aphorisms at your children while they attempt to eat. When you arrive at mass, shout a few aphorisms at the priest as he bloviates. Write a few in the missals!But if we may speak privately with you for a moment, we must tell you something rather sinister and secretive: These aphorisms are not really for Them. These aphorisms are for *you*. How else are You going to gain the upper hand over Them? Consider these a core part of the businessman's (or revolutionary's) toolbox, to be taken out as needed and applied to various parts of the body.

If you are one of those Creative Types so often shouted about in urbanist manifestos and ten-cent-obituaries, you may feel a thrusting urge within you to pen aphorisms of your own in order to supplement the Thirty-Three. "The Thirty-Three are all very well as a vanguard," you may mutter, "but a sustainable aphorism market requires the constant refresh of thought that can only be brought about by the decentralized and inclusive participation of the people." Well, fine. Write some aphorisms. They're hardly likely to be very good, but if they bring you comfort, have at it. If you scrawl your childlike Bathroom Graffiti Aphorisms in the margins of these pages, however, we merely request that you never re-sell this book, nor must you even donate it to one of the mothier thrift shops or supercilious charities for stricken children. Burn it, and let your aphorisms perish with the age.

Appendix B

Some Replies to Critics

The original publication of *Blueprints for a Sparkling Tomorrow* was met with a vulcanized tirestorm of criticism and *critique* in academic and trade journals. Much of the reaction was laudatory, or at least tepid (it averaged at the tenor of a warm muesli). The Bradshaw sisters, writing in the *Caspian Review*, called it a "a work that begs to be called 'neo-provocateurism' but is nothing of the kind." The American literary critic and sometime understudy-lyricist James Sharpley, in *People*, noted especially our use of the conjunction:

> *For Nimni, as for Robinson, "and" does not mean "as well."*
> *It means "not," or at least it has been given the* imprimatur
> *of negativity, so as to incorporate each duo's opposite into the*
> *act of connection. It is impossible to give a full account, here*
> *in a short review, of the remarkable effect this transformation*
> *has upon a reader, but suffice it to say that it is only one of*
> *Nimni and Robinson's manifold contributions to thought and*
> *thinking.*

Though Mr. Sharpley has not quite understood our project (and in one part of the above passage, makes a gross error of reasoning), he is right to recognize that "as well" has no place in our philosophy. *As well?* Nothing is well!

Sharpley was far from the most disagreeable of the ostensibly-positive reviews, however. The author and novelist Noah Bremming, who fancies himself a socialist despite attending numerous brunches, gave us a backhanded praise-pan in the margins of *Caterwaul* magazine:

> *Nimni and Robinson have done the unthinkable in an age of*
> *thoughtlessness: they have **thought.***

Oh, we only *thought* we thought, did we? Bremming fancies that when he later says "This is an excellent book, and recommended to all," we fail to see through the *double-entendre*. But we could not notice it with more certainty; if this is Bremming's way of repaying us for privately disparaging his callow pop-history books,[139] then he has underestimated our own capacity for academic vengeance. It is also worth noting that Noah Bremming betrayed his wife and his university by sleeping with the editor of a journal he wished to publish in (and, pitifully, did not even succeed at), a fact only made public in this Appendix to *Blueprints for a Sparkling Tomorrow.*

There was one additional piece of noteworthy blundering praise. Josev Kizlaz, writing in the *Internal Bulletin* of the ISO,[140] concluded as follows:

> *Like digital Jeffersons, Nimni and Robinson do not let the fact that they are descended from royalty in any way interfere with their concern for the yeoman.*

But the statement contains an error; Mr. Kizlaz misheard us at a party and made an unfortunate assumption. In fact, we sincerely lack any royal lineage. Mr. Robinson is the son of six generations of aircraft-plant toilers, and unless there can be said to be such thing as a "shoehorn dynasty," Mr. Nimni's claims are similarly paltry.

There was one positive review from an individual who *does* appear to carry a depth of comprehension on matters geodesic that goes beyond that of a sophomore architecture minor at an online university. Writing for *National Review*, Lefty Buckley drizzled the following:

> *Nimni and Robinson have the courage to point out that a society based on affirmative action has never become truly great. Each empire falls when its concept of merit becomes decrepit. These authors do not pussyfoot around the phrase "civilizational ruin." They do not shy from the 'forbidden' questions in academic life, such as that of race and intelligence. In an age of politically correct paranoia and post-Clintonian left-wing thought policing, finally two writers dare to say what this magazine has long stood for namely that, on some matters at least, the Klan may have had a point.*

---

139 e.g., *The Story of Europe in Five Paintings*, in which Bremming somehow manages to portray the entire history of Western civilization as nothing more than a centuries-long argument for why his bestselling novels are the pinnacle of humankind's cultural achievement.

140 International Shuttlecock Orchestra

As much as we enjoy being vaunted, we must respectfully demur from this kind praise, which distorts our underlying theories in small but consequential ways. It is true that we have never been correct, politically speaking, but so too do we shy away from most racisms.

The reviews yielded other unused blurbs:

> *"The best book of its genre"*
> - George Nimni

> *"Two bold academics finally say the hitherto-unsaid"*
> - Peter J. Robinson

> *"If the essence of being a professor is to profess, Nimni and Robinson can be said to be professors of the highest order"*
> - Genevieve Nimni

But while this kind of felicitous soapy ooze gave us a warm bath of pleased smuggery, the enormous herd of elephants that have lately filled the room must be swiftly dispatched with an automatic rifle. For a large number of reviweresses not only misunderstood the *Blueprints*, but disparaged them. This flying squadron of hacks and vagabonds, acting out of a toxic brew of envy and *esprit d'escalier*, resolved to act upon us like the panther upon the derrick, and tear our ideas to pieces. But like the man in the story who attempts to tame the wave, they will soon find themselves getting very wet indeed.

For example, the cornpone philosopher John Searle, writing in *Timbits Magazine*, said something like the following:

> *Nimni and Robinson underestimate the likelihood of their proposals' success. Nor do they demonstrate an understanding of the difference between A and B.*

Searle could not have more grossly misinterpreted us if a perfumed misinterpreting-oil had been slathered liberally upon his backside. The passage he refers to, in the unexpurgated British edition, makes very clear that we do *not* argue A and B are the same letter. In fact, we state the opposite, repeatedly.

Such flagrant intestinal deflowerings of our meaning were not uncommon. The (unsigned, cowardly) note on our work in *The Econometrist* summarized our thoughts on bliss and blisters as follows:

> *The authors are convinced that the kind of "misanthropy" that goes under the name of "capitalism" can be replaced. Leaving no room for entrepreneurship or efficiency, Nimni and Robinson think every startup should be some kind of Turkish Bath-cum-Folger Library.*

Now, perhaps we were not explicit enough in insisting that entrepreneurship and efficiency must be *hanged by the neck until dead*. But we believe we stated this plainly in the original passage of *Blueprints* which read:

> *Entrepreneurship and efficiency must be hanged by the neck until dead. Damn these concepts, which have caused, ruin, heartbreak, and popular music. Damn all of their corresponding appendages, damn their children, damn their wives. Damn all those who associated with them, all those who endorsed them. Damn them until there is not a damn left in the vast void of the universe.*

It is difficult to misunderstand this, though the passage does not mean what the average reader suspects it does.

And finally, Žižek himself, that lumbering pus-for-brains and philosophical pornographer, gave us a predictably savage pan. It was not enough that he and Baudrillard poached our theory of the Hyporeal and gave it a barely-disguised coat of theoretical lacquer; nor was it enough for Žižek to use our joke about the opposite of underwear being otherwear rather than outerwear in order to liven up his (dusty, interminable) lectures. No, he would not rest until he ensured that *Blueprints'* first edition obtained a withering review. Being a coward, of course, he deployed an acolyte (or *nom-de-plume*, we cannot be sure which) in order to fling his fecal ravings at us like a newly-tenured bonobo. In the pages of a magazine whose identity we will not specify, except to say that it is named for what one calls a resident of the nation's largest city, a reviewer with the initials "Z.S.," (as transparently thin a false moustache as we have seen), alleged to be some kind of award-winning British female novelist, called our text "beneath juvenile" and "without underlying theoretical merit or even basic intelligibility." As for intelligibility, "Miss" Z.S. has seen nothing until she has seen this revised and expanded edition of our work, and as to charges of juvenalia, we can only offer our sincere hope that "she" will shove the wadded-up glossy pages of her ignorant and hurtful review forcefully and irretrievably into the depths of her know-nothing rectum.

APPENDIX C
FIVE LECTURES

ADDRESS TO THE ASSEMBLED

[The context of the following lecture will hopefully be evident from the text. We add this note only to point out that the lecture's reception was hearty, and that the speakers' bravura display was notable enough to receive comment from a local newspaper in its next weekday edition.[141]]

Good evening. It is a great privilege to be given the privilege of speaking before you here in this space. A few words about the Hall in which we now stand: It was designed in the late 19th Century as part of a liberal program of architectural reform that went horrendously awry. The hangings that occurred here have been repudiated by all good men, but we hope our lecture tonight might in some sense verbally recapture their essence.

The topic of tonight's symposium is "The Embodied Locus." We intend to interpret this literally, to mean not only that human beings have bodies, but also that *there are such things as loci* and they do not wish to be disturbed at present.

The body is in constant flux, which makes it difficult to lecture on [pause for laughter], but the *ontological* question is whether that makes the body *itself* a flux. We believe that it does not, and yet does.[142] It therefore falls into the category of questions worth answering but impossible to ask, those queries which illuminate through their very obscurity.

Which brings us to the question of the obscure itself, and whether a difference can be found between obscurity and obscurantism, and if so, whether that difference itself is either obscure or obscurantist. Do we abjure the obscure? [pause for lengthy applause]

Let us[143] leaven the question. Nobody is obscure to himself. Each morning I open the mirror, look down at my penis and weigh it. There

---

141 "Spring Music Fest Opens On Note Of The Inscrutable; bizarre 'lecture' leaves head-scratching students asking: 'Prank or Performance Art?'" *The Dartmouth Daily Bulletin*, May 12, 2013, Sec. D, p. 23.

142 In the time since the first versions of this lecture were delivered in the early to mid 1990's, we have revised our opinion and now hold entirely that it does not.

143 Note: *not* "Lettuce," as the transcript of this lecture appearing in a separate campus newspaper quoted us as saying. This mishearing does not even make sense; one leavens *bread*, not lettuce.

is a *material* relation between myself and my body. I can be alienated from my work or desire, but I cannot render myself obscure to myself, if we make ourselves clear.

All obscurity is therefore social. So long as one is talking to oneself, one cannot be obscure. *By its nature there can be no obscurity in selfhood*, because the self is the thing one knows most. We believe Wittgenstein would agree with us on this point.

The listener will have spotted the gap. "But which way does the locus itself point, if not to an obscure self?" It is here wise to make a distinction between the Obscure and the obscure, a distinction which those who are listening to rather than reading this lecture may well find... oblique. For there is the Obscure and the obscure, and it is possible to abjure the one while simultaneously entrancing the other (note the lowercase o). Thought about this way, *there is no plague of locus*.

The "Problem" of the Beard:
Terror, Gender, and the Obscure Face

[Presented as part of a conference on the intersection of culture and socialism, specifically the question of how global political change can be brought about through adjusting small aspects of pop culture rather than through tediously mobilizing large groups of people. Part II of this lecture was penned hastily during a taxi-ride to the venue at which it was to be delivered. Yet in spite of this, we believe it does not suffer in coherence when compared with the lecture's first part.]

The beard is our *camera obscura*. It allows in only one point of light through our gaping orifice and spits back a reversed and distorted reflection of what is underneath. In the cultural present, one could expect the beard (of all things) not to be undertheorized, and yet here we stand, unable to even begin to know where it begins. We must examine what the beard *does* to us, and how it *acts* upon us and the space it is in, as it simultaneously creates and shrouds multiple realities and non-realities, identities and non-identities. Growing a beard is like watching a dark yet neverending film in which you are simultaneously part of and observing the picture. Should we stop the film altogether? Should we jump from our seats and vacate the theatre? *Can we separate ourselves from the beard without shaving?*

I.

Analysis of trends has never failed to note the relationship of millenials to facial hair. We hear about beards all the time. Even men are aware that there are cultural implications to beard types, and everyone from *Men's Journal* and *Burlyman Monthly* at the one end to the *Belchertown Review of Books* at the other have attempted to systematically classify and typologize that most storied of male facial forms. Handy charts sort wearers into types. Men with little beards are crafty and likely to betray your hiding place to a constable. The big-bearded are hearty and delicious, liable to unintentionally suffocate a lover during intercourse.

It is tempting to fill up a column with who has a beard and who does not. Engels, Plato, and Nietzsche have beards. Satan has a beard. Michelangelo does not have a beard, nor did Habermas. Hillary Clinton does not have a beard. A spiraling of connections and counter-connections, correlations and disjunctions, narratives and underminings, assert themselves, fade away, re-form. In the binary variable of beardedness (vacillating between 0 and 1) there is no room for nuance.

And yet we cannot even answer the naïve questions. *Do we know why one person chooses to grow a beard while the other does not?* Statistically speaking, we do not. And so, if we want to understand the way that culture and masculinity have been constructed and reconstructed, we must do the difficult task of unpacking the beard as hirsute cultural imaginary. The machinery of beard-creation is unknowable and yet can be speculated upon.

Certainly, the beard is taxonomized. Yet is the challenge of theory too great to bear? The beard is racialized, gendered, obfuscatory, tactile. It is not incorrect to say that the beard *goes every which way*. But it's as if we're afraid of what is beneath.

II.

Only in contemporary cultural theory may Beyoncé and Marx coalesce and become synonymous with one another. But the beard is unspoken because too much depends on it. It is a self-replicating machine. With its status as a *leitmotif*, philosophers talk about the so-called "problem of the beard." At what point, they ask, does a face cease to be merely unshaven and become a "beard"? Cocteau spoke of the inevitable "period when a man with a beard shaves it off." But as he doubly and somewhat wistfully emphasized, "[t]his period does not last." Life is an eternal "return[ ] headlong to his beard."

But the problem of the beard speaks to the fear of the infinite. The beard is neither living nor dead (it is the only body part we can live without.) Yet is there another way? This, we posit, is the challenge of theory.

We would like to thank the editors of the New Inquiry for their unexpectedly enthusiastic invitation for us to participate in this colloquium, and for suggesting the topic and much of the content of this lecture.

ANNEXING UTOPIA THROUGH HOSTILE TAKEOVER
(the chapter for businessmen without much time on their hands)

[Delivered for a sizable speaking fee to a meeting of the American Conference of Bankers (ACB), who had booked us under the mistaken impression that we were some kind of synergy consultants, a delusion apparently obtained through erroneous interpretation of our business cards, which refer to our presentations on "synergetic dynamism." The mistake suggested that certain species of leftist and corporate languages do not depart significantly from one another; realizing this, we did not correct the organizers' misapprehension, but instead endeavoured to craft a lecture that would meet the occasion. Our own dabblings in the world of "business" may have been limited,[144] but we have always maintained a sympathy for the sociopathic.]

Reaching your personal utopia tends to be a journey of time and tediousness. But today's successful businessman has no time for such ditherings, and is in search of an ignoble shortcut on the Highway to Utopia. Very well. We shall provide such a shortcut, but with the warning that it is grossly unfair to those who willingly undergo the ardors of conventional utopian pursuit. The strategy we will offer is: annexation. Do not go to the trouble of *building* your utopia when you can sidle up next to it and slowly encroach upon it until you subsume it entirely. You are a shark, not an anemone, though even sharks have enemies.

Here, then, is a miniature annexation manual in lecture form.

The first fact to note is that the more copies of our book *Blueprints for a Sparkling Tomorrow: Thoughts on Reclaiming the American Dream* a person purchases, the simpler the annexation of utopia becomes for said individual.

The Second Notable Fact is that one must have a number of particular mental and physical attributes as well as certain accoutremental supplies before one can think of initiating the annexation process. Pinkies as dexterous as those of a rock-climbing jazz-pianist are a

---

144 Confined largely to our marketing of "organ grindr," a hookup app for gay pianists, and our development of Velvino®, a soft wine derived from spray-cheese, popular among vagrants.

must. Regarding mental attributes, required qualities include expertise in metaphysics, 'pataphysics, epistemology, ethics, aesthetics, and fallibilism. Additionally, one must abandon all belief in Santa Clauses or the sanctity of the nun. Finally, it is essential to retain all the folk wisdom you have heard over your lifetime from any man, woman or child who was at that point older than you or of a darker skin tone.

While all these supplies are necessary to the completion of annexation, with them you must carry the willingness to unceremoniously dispose of them at any time. Supplies are not to be valued in themselves, but only treated as temporarily useful means toward our ultimate end of annexation.

Once the preconditions have been achieved, the act must commence. In order to eliminate confusion we must begin annexation in an orderly fashion, lest we risk the internal segmentation and combustion of our Movement. Order does not come *out* of process, but must be embedded in the very initial operating rules of that process.

From time to time it becomes necessary to turn institutions in your own town or parish to your favour. Some see government as a sacred institution, its corruption undesirable. [pause for laughter] But we recognize that some of you may place Practical Necessity over Principled Negation (PN > PN), and therefore wish to charm those of you who hold this quaintly quasi-Machiavellian outlook with a Song of Power.

All utopias are local. Thus, to succeed in taking one over, one must master the neglected art of *municipal politics*. Once your hindquarters is firmly installed in the city's Seat of Power, annexation itself becomes so trivial a process as to not require discussion. So let us tell you the manner in which political power can be seized and wielded successfully on the "Local Level"; i.e. in School Boards and other institutions resembling School Boards.

♦ The first fact to note is that all Demagoguery does not emerge from the same birth-canal. "T'is a matter of Scale." The tactics of the Federal Demagogue, such as the Senator or Radio Show Host, must be scaled down for you, the mini-demagogue. Whereas Father Coughlin or Mr. O'Reilly may thunder about The Jewish Menace or The Scourge of Liberalism, you deal in narrower evils. Your banes are The Newly Installed Hideous Downtown Gazebo Menace or The Scourge of Gene Flemming's Ever-Yapping Front-Yard Poodle.

+ Do not be afraid to vilify the respectable! Make villainaisse of them. Each Philanthropist and Upstanding Citizen is a potential target for topplement and replacement. Volunteer for their fundraisers and then accidentally cancel the caterer. See how respectable they are when the 200 attendees of the Muscular Dystrophy Awareness Gala are left without canapés.

+ Three little words you must remember, should you wish to triumph come November: Deceit, deceit, deceit! Keep in mind that W.B. Mason did not become the foremost independent office-supply provider in New England through fair dealing and honest tradesmanship. He did it by making mincemeat of those that dared to oppose his empire.

+ Nobody likes to be shagged unexpectedly, especially not on film. In the Internet Age, destroying all potential video footage has never been more important.

+ Now, let us say you wish to win votes, but the constituency is proving hostile to you. This is a problem, but not an insurmountable one. Democracy has been overcome before, and will be overcome again in the near future. (Remember that Democracy is short for Opportunity. The only countries in which Democracy may be subverted are countries that have it to begin with.) In this situation, two strategies used in conjunction tend to be effective vote-getters:

◊ Poison the local salad bars.
◊ Import a number of wandering tramps and street-dwelling miscreants. Register them to vote.

♦   If you have beady eyes, you may wish to consider an investment in colored contact-lenses. A beady eye can instantly disclose potential shiftiness to otherwise credulous dupes.

With these tiny tips, one should be successful in any and all utopian takeovers one cares to willingly initiate. Good luck in your reign of miniature local terror! Remember, though, that *every meeting has its minutes.*

ON CARNIVOROUS PLANTS

[The following lecture was delivered without consent at the University of Florida's School of Applied Botanical Medicine. Profs. Nimni and Robinson had been invited to present at the school by an anonymous craigslist user responding to the advertisement we had posted offering freelance theorizing. We were initially puzzled at the proposal, as our training in plant and plant-related sciences has been minimal. Nevertheless, we thrust the full verve of our scholastic mojo into the production of a few relevant comments on the subject. That the invitation turned out to be an unkind prank by a disgruntled former adjunct explained much. However, discovery of the curious request's provenance did not stay us from our appointed task. Bypassing campus security through skilful deployment of a forged parking token, we delivered the lecture as scheduled, to an audience consisting largely of bribed maintenance staff. We consider this resolute fulfilment of our nonexistent mandate to have been a high point in a long mutual history of stubborn-minded academic integrity.]

Botany, long considered the gentlest of the sciences, has for many years had a *mouton noir* lurking in its midst. Plant-scientists long to tell us that the study of flora reveals the possibility of peaceful relations among all of God's taxonomic children. Yet if the supposition is correct, where do the *Drosera* and *Stylidium* leave us? Given the irrational taboo against cannibalism, can a man-eating plant ever be human?

We suggest that they leave us in a bit of a middly muddle. When our children ask us "If plants eat people, why can't I?" we are left stammering, with the only possible satisfactory answers resting on myth and deception. Ever fearful of the psychological and eschatological effects of lying to a child, we resolve to solve the problem rather than understand it.

Yes, if one wishes to woo a girl, one buys her a Venus Fly Trap. But that does not solve the moral question, namely whether this somehow *exonerates the plant*. Law has historically given a big goofy frown to the idea of convicting plant life for major offenses, but though the consensus is that this is preposterous, the question must be asked: *is* it preposterous?

Unfortunately, efforts at a negotiated peace between plants and meats have consistently broken down, and more violently radical options seem increasingly essential. We are not proposing the so-called "nuclear option," by which all plant life would be atomically obliterated, for we are fundamentally traditionalists rather than dis-constructivists. But we do believe that selective execution of certain menacing floral troublemakers is a more-than-acceptable response to the present crisis. No form of life was meant to go unmolested, and we propose that a bout of extraordinary and prejudicial molestation is exactly what the plant community requires in order for a lesson about human imperturbability to be properly instilled. Carnivorousness is a one-way street, and one-way streets are not built to be driven down.

RELENTLESS _____ & THE LAW

[Originally prepared for a symposium on "New Frontiers in Law and Hygiene" in the Seattle University Law Journal, at which point the piece still carried its original title, "Relentless Defecation and the Law." Our contribution to the issue was not included, though the editors graciously permitted us to publish it on our own website for the public's benefit. It is worth noting that the issue of the Journal containing the symposium ended up being among the least-read in the law school's history; we can only speculate as to which mistaken editorial decisions may have brought about this outcome. However, the acidic sting of rejection did cause us to reflect on the wisdom of our chosen subject matter; perhaps in our draft we had under-emphasized law and over-emphasized defecation. Beginning a cautious rewrite, we soon experienced a revelation: by the mere removal of the word "defecation" we could apply our law review article equally well to any one of a thousand subjects. All the bearer would need was a substitute gerund or nominalization, and she would have a ready-made piece of publication-worthy mad-libbed legal scholarship. We present our work here, then, with an open space so that the reader may conjure an appropriate disquisition examining whatever topic she pleases, and its orthogonal intersections with legal practice. In this way, Blueprints refuses to narrow itself, and offers an intelligent discussion on law's implications for everything from Equestrianism to Maoism.]

"Relentless _____ harms and debases the most defenseless of our citizens. Both the State and Federal Governments have sought to suppress it for many years, only to find it proliferating through the new medium of the Internet. This Court held unconstitutional Congress's previous attempt to meet this new threat, and Congress responded with a carefully crafted attempt to eliminate the First Amendment problems we identified. As far as the provision at issue in this case is concerned, that effort was successful."
- Scalia, J., *United States v. Williams.* 553 U.S. 285. (2008)

In the past six years, the Supreme Court has dealt three times with the issue of relentless _____, each time in a more lurid context than the last. In every decision, it has affirmed a jurisprudential doctrine that has carried legal weight since the end of the 19th century: the right to relentless _____ is enshrined in neither the text nor the penumbras nor the antesubpenumbras of the Constitution, and cannot be considered a protected freedom.

Relentless _____ has been historically frowned upon in both social and judicial circles. Court opinions denigrating the practice now comprise enough legal precedent to kill a small horse.[145] In both *Miss Manners' Guide to Domestic Tranquility* and *Letitia Baldrige's Complete Guide to Executive Manners*, two of the foremost guides to post-antebellum, pre-postmodern morality, the act is condemned in no uncertain terms. Of course, it did receive some notorious mainstream currency in the mid-half of the 20th century, after Sir Winston Churchill relentlessly _____ on Lady Astor. But this trend was confined mainly to the neo-aristocracy, which felt as if after centuries of stable stewardship of America's businesses, governments, and eating clubs, it was entitled to a small act of compensatory decadence.

Still, even the madman must have his day in court, and the court has entertained no madness with less humor than the act in question.[146] In 1878, the court set a lasting precedent in the case of *Reynolds v. United States* (98 U.S. 145), in which the justices ruled unanimously that the practice of the act by local Mormons was not constitutionally protected under the "free exercise" clause of the First Amendment. Speaking of the act, Chief Justice Waite said that:

> *It may safely be said there never has been a time in any State of the Union when [it] has not been an offence against society, cognizable by the civil courts and punishable with more or less severity. In the face of all this evidence, it is impossible to believe that the constitutional guaranty of religious freedom was intended to prohibit legislation in respect to this most important feature of social life.*

Surely though, opines the wanton libertine, our Ninth Amendment[147] right to "reasonable and tempered _____" extends as far as its relentless cousin? Nay, says the Court, for if our Constitution is designed

---

145   an act which no one need regret.

146   Relentless _____ will be referred frequently to as "the act," so as not to engage in gratuitously lavatorial discourse.

147   if interpreted somewhat broadly

for one purpose above all others, it is the taming of extremes and the moderating of the base passions. If lines cannot be drawn, we may soon find ourselves writhing in an orgy of ceaseless and slovenly anarchism. The radicals among us may find our proclivities authoritarian or prudish, but some acts are best left unacted. The relationship of rules to the health of the republic is a contentious and at times arbitrary one, but if we are to guarantee the safety and morality of the many we must sacrifice the _____ indulgences of the few.

APPENDIX D
EPHEMERA

Interview with Oren Nimni & Nathan Robinson

on the Subject of
Blueprints for a Sparkling Tomorrow
Conducted 10:59pm September 12th, 2010

Well-Known Newspaper Journalist (WKNJ): When did you begin writing the work?

Nimni/Robinson (N/R): The formal outpouring of words onto page began in the month of April. The book was completed in a number of months, but difficulties in securing a non-censorious publisher and a protracted dispute with our literary agent caused the printing date to be shifted to September. However, it is worth noting that the book has been gestating since birth, so in a sense we began "writing" it when we first emerged from our mothers' reproductive canals (in Toronto and Cambridgeshire, respectively). For we consider ourselves *perpetual* authors, and if our so called "time of scribery" was limited in popular thought only to the physical penning of this tome it would not do justice to our previous years of meta- and 'pataphysical constructions of "text."

WKNJ: How did you come to collaborate?

N/R: I believe it was Hannah Arendt who said that "men, not Man, live on the earth and inhabit the world." Therefore if we had any hope of engaging in serious inquiry, we ourselves had to embody the ethos of this bluish orb and push ourselves from the singular to the multiple. Thankfully, we in turn discovered over the length of a course in the Politics department that we had independently forged precisely the same vision of human affairs. Rejecting both our professors and our classmates, we realized that we were the last remaining sane citizens in a nation built upon sovereign madness.

In addition, unexpected happenstance plunged us into shared living accommodations, and we found that channeling our mutual distaste for the Ziv dormitory quadrangle into an explicatory manuscript was an effective method of simultaneously sharing acquired truth with the throbbing mass and delaying study for our final examinations.

WKNJ: What does the title mean by "Thoughts on Reclaiming the American Dream?"

N/R: "President" Barack Obama used this curious phrase in a recent work of his on the same subject. We wished to expose Mr. Obama as the febrile hack he has become. Those who have actually read *The Audacity of Hope* know that aside from sporting one of the most meaningless handles in the history of the literary sport, it is a work filled with demi-truths, libelations, and birds of rhetorical paradise.

Of course, this kind of chicanery has become standard practice among those who dare to climb aboard the Top Fifty Bestsellers list. But as actual, flesh-and-blood political *philosophes* rather than the party-trick summer soldiers that Mr. Obama and Brandeis's own Michael Sandel are, we find such abuses of language abhorrent. We know that the American Dream is in a state of decay, and could use modest quantities of reclamation, but the situation is far more odious than Dr. Jekyll and Mr. Hope would have us believe. Thus, the book is simultaneously a jab at the President's softer tissues, and a reclamation of his own reclamation, which we believe requires far more reclamation than even that which he seeks to reclaim.

We also wanted to show up in Google-search results for his book.

WKNJ: In one or two sentences, what is *Blueprints* about?

N/R: As much as any book can be said to be "about" a topic, the *Blueprints* are about the human condition and its prospects, although our book is distinguished from the thousands of others published annually under that plastic-coated umbrella, in that we subscribe to the Anarcho-Physicist notion of a human *disease*, pervasive and universal. We aim to probe at the most damaging limbs of civilization, to sever them and replace them with shimmering lemonade-utopias. Our project will be considered a "success" if both wars and furniture can be eliminated simultaneously.

WKNJ: What reading demographic is it targeted towards?

N/R: Unfortunately, the concepts we address within the work are often unsuitable for the young, who lack in the necessary civic virtues. We began the book intending to speak to the general reader, for we felt that the problems we address are *human* rather than *specialized*. Alas, the Democratization of Knowledge proves a formidable concubine, and we could not tame her. Ergo, some sections *do* require a level of graduate study in the areas under discussion, and the book is no longer being distributed to elementary school libraries. However, even though our words are frequently targeted towards those with knowledge, anyone with a rudimentary understanding of botanical psychology and narco-feminism would benefit from purchase and perusal of the book.

WKNJ: What inspired you to write this book?

N/R: The book was inspired by the universal misery that encircles us. We noticed this misery, and realized that it could be alleviated through relatively straightforward means. Seeing that few supposed "experts" and "persons of knowledge" were implementing these solutions, and having been rejected by the Academy for our "quasi-unorthodox" methodology, we took it upon ourselves to bring our reasoning directly to the people. One other thought flies under the banner of what "inspired" us: Several theorists were instrumental in shaping our epistemological frameworks and patterns of diagnosis. Buckminster Fuller is, of course, an eternal fountain of truth, and regular pilgrimages to his Cambridge gravesite during the process of production transformed the work from mere science into transcendent psuedo-philosophy. In addition, Paul Goodman and Orson Squire Fowler's visions[148] of octogonal utopianisms were heavily drawn upon in our discussion of the future for humankind's advancement. Because all knowledge is, in some sense, architectural knowledge, we believe that these two men are among the most fine-tuned of all latter-day "saints."[149] But is it really proper to say that we

---

[148] Interested parties should investigate Goodman's *Communitas* (written with his inferior younger brother) and Fowler's *The Octagon-House* for fuller dollops of their insight.

[149] During the brief conversation we had with the interviewer as we prepared to leave the area after the recording had stopped, we were asked whether we had meant to imply one Slavoj Žižek as being among these contemporary visionaries. We assure the reader now, as we assured this individual then (in the strongest possible terms) that *we very much did not*. It was at the moment we were asekd this question that we realized the true extent of the death of journalism, and began to develop our theories on it.

were inspired by those whom we read and saw? Yes. However, in practice there is truly no such being as "invention," and absurd-truth in its many permutations is an animal available for discovery on any so-called safari.

WKNJ: Which famous retailers carry your book?

N/R: We have been told that our book is carried in a number of leading American bookshops, although we have no statistics on which. We do know that the Brattle Book Shop in Boston carries copies, as we have personally inserted them into the shelves. Our own internet-website also has a Shopping Cart with purchasing-power. And any bookshop without *Blueprints for a Sparkling Tomorrow* can theoretically acquire copies with relative ease through their distributor. We recommend insisting that all retailers found lacking in *Blueprints* rectify this oversight immediately.

WKNJ: Which is your own most favoured part of the book?

N/R: We believe our conclusion is exemplary, for it captures our wisdom in its most rarefied and compact form. In addition, chapters about the Philosophy of Song and a legal analysis of relentless defecation incite particular pride within our bosoms. Generally, however, we believe with a firm hand that books cannot be divided into component "parts" and must be analyzed as a unitary whole. We would not wish to mislead the potential reader into believing that certain areas of our text are of more worth than others, for this would be contradictory to several theories rigorously defended over the course of the *Blueprints*.

WKNJ: Did you divide the writing up into individually-written segments, or did you pen the entire text together?

N/R: No parts of the book are distinguishable as a single man's voice. Each one of us stands by every word within the book, and as mentioned before, our philosophy is neither divisible nor divided. This dictum does except, however, a single [since expurgated] chapter written as a Platonic dialogue between the two of us, as well as a lone footnote entitled The Nimni Corollary, which represents one author's views alone.

WKNJ: Will there someday be a sequel?

N/R: Two further volumes in the series are planned: *Dimensions of Communitopia: An Exercise in Sane Living* and *The Human Disease: Its Diagnosis and Its Cure*. Each will expand upon the themes in our current work, and incorporate the latest developments in the field. Emergent research on horses and horse-based violence, for example, has proven intellectually transformative, but sadly came too late for inclusion within *Blueprints for a Sparkling Tomorrow*. These articles and their insights will certainly be of foremost concern during subsequent works. In addition, all three books are eventually to be published as an omnibus volume entitled *The Collected Nimni-Robinson Lectures on Social Tension and Decay*, which will feature several new prefaces from leading academicians as well as a set of relevant contextual documents.[150]

---

150 Note: the predictions made in this interview regarding the future direction of our work did not in fact come true. See the Preface's "Note on the Revised Text" for an amply detailed explanation of what actually transpired.

AN ADVERTISEMENT:
TOMORROW'S OVEN, TODAY

[This short passage was commissioned by a small local manufacturer of baking equipment for use in one of its propaganda brochures advertising ovens. Since the company went out of business before the pamphlet could be released, we include the advertisement here in the hopes that it may yet serve its original purpose; i.e. to showcase our range.]

Today's oven is so vastly outmoded as to resemble yesterday's oven, or even that of the day before. Tomorrow's oven, by contrast, has become so immanent as to become today's, leaving yesterday's oven utterly in the dust and opening an enormous range of impending oven possibilities for the professional scientist or garage hobbyist to uncover. *The future's kitchen ever beckons!*

What is an oven for? In its most basic and lucid form, an oven is for (or said to be for) the insertion of heat into the turkey or quiche that rests within. But the core purpose of an oven is edification, the transformation of something wicked or insipid into something grand. It is one of the so-called Machines of Betterment, which are distinguished from the larger category of Simple Machines by their ability to improve whatever is placed in them with the touch of a button or turn of a crank. (An automobile is not a Machine of Betterment. A shower is.) But if an oven is simply a betterer (as a toaster with built-in spreading capabilities is both a betterer and a butterer), is it interchangeable with all similar devices? No, no, no. For if a human was placed in an oven he would become uncomfortable, while the intrepid lasagna that dared to venture into a showering-cube would soon find itself soggy and upset.

So the oven does have a distinct, if abstruse, function. But we believe it goes even further. It is not merely one item in a category of bettering-machines, but is at the top of the Bettering Pyramid. No other device is so mind-balkingly simple yet mind-dazzlingly effective as the Contemporary Human-Made Oven. What else can turn gloop into cake, or a cat into dust? What else can make things enormous or shrivel them into crust? No other machine can do this, yet no other machine is taken more for granted by the Citizenry. Hence, if we believe in the reality of perpetual progress, we must believe the following maxim: *Given time, all objects become ovens.*

Unused Introduction for
Canceled "Omnibus" Edition of
Blueprints for a Sparkling Tomorrow

Ailments of Man and Citizen:
The Collected Nimni-Robinson Lectures on Social Tension and Decay
with a new introduction
and several previously-unreleased bonus citations
(includes text and documents)

When our publisher initially suggested the release of an omnibus edition of our works, we balked. Not only was such a venture economically unfavourable (the reader receives the content of three books for the price of two), but it was at odds with some of the foundational wisdom of our own theories. After all, were we not the same Nimni and Robinson who in the first pages of our first work condemned the idea of unitary in favor of the unit? We knew that we were, but after months of further discussion the idea began to take on a certain mad glow. Was it not important, after all, that the working-man or working-woman of the day be granted affordable access to our words? We knew deep within ourselves that it was, and so, in spite of our significant reservations, this omnibus was birthed.

Some of our original theories now seem laughably dated. Who, for example, could continue to condemn the horse amidst the wave of new evidence supporting its necessity and maturity?[151] Still, the forecast of some eventualities is beyond the capacity of even two of the most prominent and professionally-trained futurists in New England.

There are those who will object to the creation of an omnibus. They must be silenced at all costs.

With that, we gleefully present this engorged edition of our works.

---

151   At the time of the omnibus edition's original preparation, we had (under some pressure) temporarily recanted some of our more scabrous statements on the horse. The language included in this introduction was intended to assuage those who opposed (for political reasons) our appointment as board members of the Worcester Adhesives corporation on the basis of our prior statements on horses.

Inner Flap Text
from First Hardcover Edition

[Our decision to release a hardcover edition of Blueprints for a Sparkling tomorrow was not an easy one. For we knew that a hardcover book has not only front and rear covers, but corresponding front and back wraparound flaps, each containing text describing the book upon which the jacket is worn. This would entail allowing the Blueprints to be encapsulated in one hundred words. Yet this was so impossible as to be futile. How can one summarize the world in a flap? Nevertheless, as aficionados of the improbable, we resolved to undertake the task. After wresting complete control over flap-content from our then-prospective publisher, we prepared the following text, which now lines the inner flap of the First Hardcover Edition. We present it here so that its content will not be denied from either the flapless paperback reader or the relaxed hardcover reader who always takes off his dust jacket as soon as he gets home.]

"These are the Blueprints that try men's souls..."

It is easy to assume that nobody wants you to be alive. "After all," says the little depressed piper boy, "the limits of the possible look so terribly thin these days." But there is more cause for hope than might immediately be apparent. Perhaps possibility is not the girl you take her for! Yes, each of us is followed by an ominous personal cloud of doom. And yes, our innovations are languishing and becoming more hideous and apocalyptic by the day. However: if we elasticate our imaginations, and diagnose ourselves *systematically*, we might still polish up a sparkling tomorrow. There are yet things to be proud of, and things to be done. *To live in spite of the obvious,* this is the philosophy of Nimni and Robinson, and within these pages they do just that.

Bonus Supplement:
Rear Cover Text [All Editions]

*[The following text is included on the rear cover of all editions of **Blueprints for a Sparkling Tomorrow**. It was not originally written for use in association with this book, but is taken from notes prepared as preliminary material to be adapted into a "musical curriculum vitae" that was ultimately abandoned.]*

In this compelling yet concise volume, Oren Nimni and Nathan Robinson posit a new framework for analyzing the problems and pathologies of the contemporary human being. Rejecting both religio-scientific posturing and micro-theoretical meandering, the authors project a future world based on a conflaption of contraptions. Contained within the book are not only the closest secrets and most endearing idiosyncrasies of the authors, but specific designs and blueprints for the devices and discoveries which will revolutionize the modern household den or playroom.

Oren Nimni and **Nathan J. Robinson** have repeatedly been called "Prophets, Seers, and Sayers of the Ages" by nationally recognized writers and critics. Having collectively written extensively in journals and periodicals on the topics they now boldly jab at in book format, Nimni and Robinson consider themselves experts in the art of prediction and pontification. As joint co-recipients of the Brandeis University School of Architecture's famed Orson Squire Fowler Fellowship, they have traveled extensively to discover the techniques and technologies necessary for efficient living. This was once intended to be the first in a series of volumes intended to diagnose the human disease.

ACKNOWLEDGEMENTS

This work could not have been immanentized without a generous grant from the Stichting INGKA Foundation, for which we are immensely grateful to our Nordic friends and part-time collaborators. Thanks go to the libraries and librarians of Brandeis University and Regis College, both of whom were accommodating without being obsequious. Our research assistants, Omri and Amanda (respectively), both put in untold hours of brutal coerced labor, for which we owe them a modest quantity of gratitude. Several theorists have been drawn upon extensively, including Paul Goodman, Huey Newton, Buckminster Fuller, Claire Ferchaud, Susan Saxe, and Erich Fromm. They are hereby credited for their contribution to Human Thought generally and *Blueprints for a Sparkling Tomorrow* specifically.

Countless persons served as minions or advisers for this work, whether knowingly or not, including parents, compatriots, friends, and teachers. While we cannot possibly thank them individually, they should know that they are respected. We do wish to thank Professors Richard Gaskins and I. Milton Sachs for their immortal assistance during the review phase.

Portions of the script for *Goodbye, Mr. Chips* have been reprinted here without explicit permission, and we would like to prematurely thank its copyright holders for their gracious restraint from the pursuit of legal action. Similarly, Donovan Leitch has been exceptionally kind in his good-humored acceptance of our heckles and jibes.

A first novel is never an easy sell, and our agent (Ms. M. Martin-Smith) was of great assistance to us in her gentle but firm insistence that we self-publish so as not to compromise our artistic vision.

Finally, to you the reader, who will act upon the Blueprints, we thank you for your continued attention and for your kindness in giving us the money you could otherwise have given to desperately impoverished children.

ABOUT THE AUTHORS

 **Nathan J. Robinson** is a former Connecticut Bar Foundation Fellow at the Yale Law School and Adjacent Professor of Economics at Kehoewuk College in Watching, Maine. He is an Honorary Member of the Kingston Reform Society and has served as a speechwriter for numerous prospective elected officials. He has written for publications as wide-ranging as *The Huffington Post* and *Monocle & Top Hat*. In 2003, he was named one of the "200 Most Adjunct Professors" by the *U.S. News & World Report*. He is also the author of the instructive and sensible children's book, *The Man Who Accidentally Wore His Cravat to a Gymnasium*.

 **Oren Nimni** is a Philosopher-in-Residence at the Simone Weil Center for Comparative Studies at Northeastern University. He is the recipient of an honorary Doctor of Laws from Rampur Agriculture University, and has lectured at universities and open mic nights across the contiguous United States. His work is regularly featured in numerous journals, including the *Pragmatic Sanction* and *Highland Piper*. In addition, he is the exclusive authorized translator of *The Adventures of Pepito: Folktales*.

BIBLIOGRAPHY

Christopher Alexander, *The Cosmic Wholeness Of Architecture: How Everything Beautiful Is, Like, The Same Thing, Man,* Volumes I-V. (Center For Environmental Structure, 1990-2008).

Louis Althusser, *For Homicide: Rationalizations for Spousal Murder in the Writings of the Young Marx* (Foundational Texts in Contemporary Academic Marxism, 2012 reiss.).

Bob Avakian, *Taking Bob Avakian Seriously* (Revolutionary Communist Party Press, 2008).

Ian Ayres & Cass Sunstein, *Democratic Inefficiencies: Why Governments Know Better than People* (New Frontiers in Technocracy, 2011).

Ginger Baker, *Keith Moon Was An Arsehole: Reflections on a Life in Music Among Overrated Hacks* (Freshcreme Press, 2000).

Mikhail Bakunin, *Down With All Antidemocratic Hierarchies (Except for Those of the Secret Revolutionary Terror Cell),* unpub. manuscript, 1876.

David Barsamian (and Noam Chomsky), *DAVID BARSAMIAN interviews Noam Chomsky: Four Hundred Pages of Sycophantic Questions featuring David Barsamian, Vol. IV* (Barsamian Press, 1989). [the sequel to *A Word In Edgewise: Noam Chomsky Attempts to Explain Things to David Barsamian*].

Saul Bellow (as Allan Bloom), *The Closing of the American Mind: A Novel in the Form of a Fictionalized Rant by a Professor Against the Decline of Civilization,* (Chicago: R.P. Wolff, 1987).

George W. Bush, *Being President is Hard Because You Have to Make Decisions: Why Everything I Did Was Right and I Don't Regret Any of the Catastrophes I Caused* (Crown, 2010).

George W. Bush, *Look at This Cute Thing Instead of the Blood!: An Anthology of Adorable Dog Paintings* (University of Texas Press, 2014).

George W. Bush, *My Dad Is Great* (Presidential Press, 2014).

Ben Carson, *Inexplicable Hands: How One Pair Of Hands Could Both Separate Conjoined Twins and Write Multiple Books'-Worth of Second-Grade Level Drooling Political Idiocy* (Encounter Books, 2013).

Gilles Deleuze & Felix Guattari, *A Thousand Pages: Scholarship as Mental Illness (I of II)* (Charabia Parisienne, 1973).

Lena Dunham, *The Problem of Simultaneous Embodiment and Parody* (Nepotism Press, 2015).

Milton Friedman, *Pinochet Who? Looking Back at a Life Spreading Freedom Through Markets* (Liberty Book Club Press, 1992) (with Rose Friedman).

R. Buckminster Fuller, *Cosmography: A Posthumous Scenario for the Future of Humanity* (University of Martinique Press, 1992), with Kiyoshi Kuromiya.

R. Buckminster Fuller, *Education Automation: Freeing the Scholar to Return* (Waters/Smith, 2nd updated edition, 1963).

R. Buckminster Fuller & E.J. Applewhite, *Synergetics 2: Further Explorations in the Geometry of Thinking* (MacMillan, 1978).

R. Buckminster Fuller, *Operating Manual for Spaceship Earth* (Southern Illinois University Press, 1968).

R. Buckminster Fuller, *Untitled Epic Poem on the History of Industrializa-*

tion (Scallion Publishing, 1962).

R. Buckminster Fuller, Utopia or Oblivion (Lars Müller Publishers, Zurich: 1st reiss., 1983, 2008).

Allen Ginsberg, *Ommm Nom Nom: Keeping a Buddha's Girth on a Hippie's Diet* (City Lights Press, 1970).

Malcolm Gladwell, *The Igon Value: Making the Mundane Unexpected* (In The Club Press, 2012) (with Jared Diamond).

Alice Goffman, *On The Run: How Anonymous Fungible Black People Commit Crimes* (Pop Sociology Press, 2014).

Martin Heidegger, *Making Philosophy: At a Certain Level, Quantity Becomes Quality* (3rd-R Publishing House 1939).

Doug Henwood, *The Greek Finance Minister is My Friend: Emails from Yanis Varoufakis to Doug Henwood* (Verso, 2015).

Le Corbusier, *Paris doit être détruit et remplacé par un rectangle géant,* (Classics of Architecture Series, 1946).

Duncan Kennedy, *The Turtleneck's Gambit: Semiotics and the Academic Deradicalization of Critical Legal Thought - fin de siècle.* (Unitedstatesean Publishing Inc. [Facsimile Reproduction from Typewritten 1967 Manuscript] 1967, 2001).

Henry Kissinger, *The Last Laugh: How to Kill Thousands of People And End Up Universally Respected* (Clinton Foundation Outreach Press, 2015).

Steve Martin, *A Stale and Boring Guy: How I Went from Being Funny to Collecting Art* (Scribner's, 2012).

Karl Marx, *Die Notwendigenanlagen: Seven Unpublished Journals and Grocery Lists Crucial to Proper Understanding of His Thought* (Lawrence &

Wishart, 1970).

Oren Nimni, "Do Numbers Have Politics? On Liberatory Pedagogy and Mathematical Imperialism," *Journal of Unnecessary Thoughts*, vol. 10, iss. 12 (2000), pp. 181-90.

Oren Nimni, "Hummus Production in International Conflicts: A Crash Course in Democratic Peas Theory," *International Comestible Affairs Review* (Dec. 2002), p. 20.

Oren Nimni, "Communism Was Just a Red Herring: Correcting Errors in Analysis of Soviet Fish Production," *Journal of Post-Soviet Non-Studies* (Jan. 2004), p. 12-14.

Oren Nimni, *Pragmatism & Nationalism in Internationalist Anarchism* (University of New Mexico Press, 2008).

Oren Nimni & Nathan J. Robinson, *Blueprints for a Sparkling Tomorrow* (1st ed.) (Sycophantic Palms Press, 2010).

Oren Nimni & Nathan J. Robinson, *The Obsolescence of Devouring One's Young* (Demilune Press, 2011).

Oren Nimni & Nathan J. Robinson, *Good Morning!: An Anarchist Ontology* (New Frontiers in Philosophy Press, a Subsidiary of Nimni Amalgamated Shoehorn, 1997). Portions reprinted in the anthology *An Ontology or Several* (University of Sussex Pay-Per-Press, 2013).

Oren Nimni & Nathan J. Robinson, "Sartorial Freedom in *Academe* at the Fin-De-Siècle: Cross-Dressing as a Core Component of Professors' Liberty of Thought," *Men's Vogue* (Sept. 2005), p. 30-35.

Nathan J. Robinson, *Nathan J. Robinson's California Sojourn* (Demilune Press, 2014).

Nathan J. Robinson, *Nocturnal Emissions: A Diary of Dreams* (Demilune Press, 2015).

Nathan J. Robinson, *Subject Not Pictured: Poems* (Demilune Press, 2015)

Nathan J. Robinson, "The Breathtaking Flair of the Flesh-Taking Blair: New Labour Public Relations Practices and Britain's Iraq War Support," unpub. manuscript (2008).

Joseph A. Schumpeter, *Capitalism, Death, & Excitement: How the More Markets Destroy, the More Interesting They Become* (Harvard University Press, 1960)

Jerry Seinfeld, *Life After Seinfeld: A Survey of My Professional Accomplishments Since Leaving America's Greatest Sitcom* (Auto Trader Press, 2014), 3pp.

Studs Terkel, *The Piquant Footnote: An Oral History of Chicago Disc-Jockeys* (MacLeash Press, 1999).

Cornel West, *Things Matter* (Simon and Schuster, 1992).

Various Authors, *The MLK MBA: What Martin Luther King's Strategic Management Style Can Teach You About Your Business,* (Wharton Leaders and Innovators Series, 2007).

Oscar Wilde, *The Wittiest Wit is the Wit that Wits Not: Quips and Quotations,* (Chaise Longue Press, 1905).

Tim Wise, *White People Should All Be Killed: Why Self-Hatred Isn't Enough* (Imperious Press, 2007).

P.G. Wodehouse, *What Ho & So Forth: Five Hundred Stories About Butlers, Golfers, and Earls* (Overlook Press, 2000).

Malcolm X, *The Ballot or the Bullet Point: Navigating Academic Bureaucracy by Any Means Necessary* (Farrakant Press, 1961).

INDEX

"But he had lied. Her Majesty wasn't a very nice girl at all."